THE CITY

THE PECKHAM EXPERIMENT

SOCIOLOGY OF THE CITY

THE PECKHAM EXPERIMENT

INNES H. PEARSE & LUCY H. CROCKER

Routledge
Taylor & Francis Group
LONDON AND NEW YORK

First published in 1943

This edition published in 2007
Routledge
2 Park Square, Milton Park, Abingdon, Oxon, OX14 4RN

Routledge is an imprint of Taylor & Francis Group, an informa business

Printed and bound in Great Britain

British Library Cataloguing in Publication Data
A CIP catalogue record for this book
is available from the British Library

The Peckham Experiment
ISBN10: 0-415-41749-X (volume)
ISBN10: 0-415-41931-X (subset)
ISBN10: 0-415-41318-4 (set)

ISBN13: 978-0-415-41749-5 (volume)
ISBN13: 978-0-415-41931-4 (subset)
ISBN13: 978-0-415-41318-3 (set)

Routledge Library Editions: The City

A Sir Halley Stewart Trust Publication

THE PECKHAM EXPERIMENT

a study in

The Living Structure of Society

by

INNES H. PEARSE, M.D.
LUCY H. CROCKER, B.Sc.

Published for
the Sir Halley Stewart Trust
by GEORGE ALLEN AND UNWIN, LTD
40 MUSEUM STREET, W.C.

FIRST PUBLISHED IN 1943
SECOND IMPRESSION 1943
THIRD IMPRESSION 1944
FOURTH IMPRESSION 1944
FIFTH IMPRESSION 1944
SIXTH IMPRESSION 1947

Printed in Great Britain by
THE NORTHAMPTONSHIRE PRINTING & PUBLISHING CO., LTD.,
at their OFFICES AT RUSHDEN.

FOREWORD—1947.

Any scientific worker hates to have his experiment interfered with. During the first world war a well intentioned head-mistress held up her hands in horror when she found a bunsen burner alight on each bench in the school laboratory, and there was dismay and indignation in the class as she sailed round patriotically turning them all out—and ruining the experiments. So the second world war came to Peckham—and turned out the lights. But this time it was no schoolgirls' hackneyed experiment but the most long term of all experiments—the biology of humanity—its unit of time a generation, its crucible of space, the home. Nothing less than the supreme expression of health—parenthood, the family and the home—was the subject of our experiment.

That the experiment could be ill served by such extinction, there can be no doubt. The days of hypothesis and mere theory were past. The building—the biologists' laboratory—was there; the terms of the experiment were set; all that was needed was time. Time for the families to congregate and to function—time for the scientist to understand what he saw.

This book is evidence of how greatly time was needed. Throughout, in tentative and enthusiastic language it describes what the scientist had begun to find. In all the departments of human activity legitimate within its walls, observations were accumulating, and principles were being elaborated. It was a thrilling experiment, gathering momentum with every year of its life, and presenting after only five years a unique richness of community life that was a revelation to all who were privileged to share in it—even to see it.

The study of health, or "wholeness," is not, as in common parlance and practice it is too often assumed to be, the study of disease, whether its cure or its prevention—which is a negative approach. It is the study of a positive entity, including, among other things, the study of vitality—its origin and flow through society. So in 1935, when the Centre, designed and built as a biologists' laboratory, was first opened, we set out to look for evidence of an acceleration of such a flow through the families gathering there. Five years demonstrated clearly that we had struck a potent source of the quality of which we were in search. That time also sufficed to confirm our earlier suspicion that there was as yet no gauge or means of measurement. Clinical criteria were hopelessly inadequate for the assessment of health.

Now the Centre starts again, with those first slow years of early growth to go through once more.

But with this difference. To all sober-minded persons it is now clear—as it has never been before—that at this juncture in civilisation man must find a new path, or perish. There are poets, scientists and many many ordinary men and women who have a glimmering of the direction in which we must move. But the so-called "planning" so much in vogue is too often no more than piecemeal correction of the mistakes of the past; a negative approach to the separate disorders of a disintegrating society. If in our planning we confirm this disintegration by the admittedly convenient and even logical method of segregation—into sex, age, wage, work and class groups—we are then faced with the necessity to administer each separate section so that together they may work as a whole—an administrative dilemma of growing dimensions facing all nations.

If, on the other hand, we keep society in its own "parcels" of wholeness, that is to say in families, then out of their natural growth and function a social whole or community will arise spontaneously, that is to say of *its own inherent vitality*.

To planned expansion there is then an alternative—*organic* growth. This is a biological phenomenon of the living world; as such it is distinct from logical planning from premises culled from the science of the physical world.

The Peckham Experiment throwing its spotlight on a fragment of society has given us a glimpse of this alternative. The Centre's forms and methods were designed for the very purpose of revealing this possibility: to enable us to search for the principles of growth and development by which ordered growth can happen.

* * * * * *

In May, 1946, the Centre re-opened. Structurally the building had suffered little, but its use as a munitions factory during the war, and loss of much glass through blast, left us in possession of what looked little better than a slum. There was no bright new building with attractive modern equipment to draw families from their much bombed, distressingly overcrowded and often leaky dwellings. Nevertheless of the 840 families (*i.e.*, roughly 3,000 individuals) making continuous use of the Centre before the war broke out, 500 families (approximately 2,000 individuals) rejoined. Ten days after the munitions firm had left they held a reunion party of 2,000 people to celebrate their return. They had their Centre back.

But things were changed—oh yes. We all of us, members and staff, somehow expected to come back to a Centre which would be the Centre that this book describes with its full and growing life moving forward in its own familiar pattern. We had forgotten that like a plant cut back to the ground, it had to grow again; forgotten that seven dark and barren years had intervened. The 'organisation' of the Centre's life was gone. Not only was there little equipment, no glass walls round the swimming bath, and the nurseries and gymnasium out of action, but there was no dramatic group with a play in rehearsal, no dance band to play on Saturday nights, few children who could swim, far less leisure and no social contacts with the out side world. All of us were seven years older, and those now adolescent, having had few opportunities during the war, were unfacultised people whose only leisure achievement was to jitterbug . . .

What would happen? Had the Centre roots? Would it—like the plant—grow out new shoots, leaves and fruits of its own inherent vitality? The year that has just passed has been one of bated interest for the observer.

Slowly, tentatively at first, then more rapidly, the social life is beginning to take shape—to 'organise' again in the terms of the biologist. By that token the Pioneer Health Centre is proving itself a living entity with the quality of health or wholeness.

Meanwhile it has not been without a far flung influence. Under the ground for seven years it has been thrusting 'suckers' into the nation's life, moulding, modifying thought and opinion, stirring the social soil with its potentialities. Two of these have already thrust above the surface—one in Coventry, now well advanced, another in Sheffield—while the buds of five or six others are astir.

So, through the dual evidence of its own renewed growth on the one hand and its power to reproduce itself on the other, we see that it is "alive." We do not hesitate to claim for the principles which underly the Centre a validity that makes them of peculiar importance at this juncture in human affairs.

That brings us to another issue. It is a testing time for original research. Is science to remain free and independent as it must be, or has the scientist to be trained to do as he is told and submit his thought and enquiry to authoratative direction and planning and cease to be original? We forget that the one and only characteristic of Science, as distinct from scientific technology, is that it *is* original and originating. It can only remain free and independent if it is supported by those who look to the future. The characteristic of the future is that it is open to exploration—and that implies faith. The characteristic of the past is that it is only open to exploitation—and that implies nothing stronger than belief in what is known. The future is the unknown—the realm which Science must continue to explore with faith.

The answer to this question is one of urgency to those concerned to see the Peckham Experiment go forward. In their anxiety to establish Peckham Health Centres in their own localities and for their own use, many of our readers and earlier supporters often forget that for their own efforts to be fruitful it is more than ever necessary that experiment and research should go on, and that the war has made it very difficult to sustain funds for research.

Further, each locality planning a Centre looks to Peckham to provide it with trained staff, so that inevitably the Peckham Health Centre has to become a Training School—an expensive and avowedly unprofitable undertaking.

To meet this dual need for research and teaching there has now been formed a National Health Trust for the study and promotion of Health. This Trust enables new Centres to be affiliated, with the Peckham Health Centre as its Research Laboratory and Training School.

There are now some 20,000 copies of the earlier editions of this book in circulation. If every reader who has been able to take from it something of value for himself and his own work, were to translate his interest into even modest support of the National Health Trust, the Peckham Health Centre could continue and expand its work without anxiety or delay. At the moment the worry of financing the experiment is so great that it is seriously interfering with the progress of the research.

AUTHORS' NOTE

This book has been written to afford an approach for the intelligent layman to the growing content of the modern science of human biology. In the coming years we are all going to discover that we must either learn to understand and live in obedience to the laws of biology—the science of living—thereby coming to live more abundantly; or that by ignoring them our misfortunes must multiply till, heaping up, they ultimately destroy man's civilisation—and even man himself.

The substance of this book has been drawn from the inspiration of and experience gathered under the Directorship of Dr. G. Scott Williamson, Halley Stewart Research Fellow, at the Pioneer Health Centre in Peckham. It is the third of a sequence of four books, of which the first two—"The Case for Action"[1] and "Biologists in Search of Material"[2]—have already appeared, and the fourth—"Science, Sanity and Synthesis"—on the scientific principles upon which the experiment is based, is yet to follow.

It was the good sense and generous co-operation of the member-families of the Centre which has made such an experiment possible at all, and if any reader finds in the pages of this book that which interests, pleases or illumines him, it is to the members of the Centre, who lived through some difficult as well as many "grand days", that thanks are due.

The work that this book represents was contributed to by every member of the Centre's staff, the observations of each having been woven into a whole by the authors—a doctor and a biologist 'curator' on that staff. Much trouble has been taken to ensure that nothing important to the issue should be omitted, and that the balance of the whole should be preserved in accordance with the facts and happenings as they occurred in the course of the experiment. For her unflagging interest and work in this connection, we wish to express our deep sense of gratitude to

[1] Pearse and Williamson (Faber & Faber, 1931).
[2] Staff Report, Pioneer Health Centre (Faber & Faber, 1938).

Miss Mary Langman, who nominally acting as amenuensis has actually contributed to every page of this book.

The money for the Peckham Experiment has been raised by a Committee of lay people, who have carried a heavy burden of anxiety in finding the necessary funds for experiment on so large a scale. The scale of the experiment was determined by the needs of *health* ; for experience has already taught us that health can only come forth from mutuality of action within a society sufficiently mixed and varied to provide for the needs of mind and spirit as well as of body. The authors trust that all those whose vision and generosity have led them to support the Pioneer Health Centre will see in this book some realisation of their hopes.

The writing of this book has been made possible by a grant from the Halley Stewart Trust, for which the authors wish to express their great appreciation.

Innes H. Pearse

Lucy H. Crocker

CONTENTS

	Historical	*page* 9
Chapter I	Living Things	15
II	Man in the Making	27
III	Basic Technique	40
	A Chapter in Photographs. The 'Centre' life	51
IV	The Health Centre	67
V	Health Overhaul	79
VI	Findings of Overhaul	93
VII	New Member-families	124
VIII	The Family Grows	135
IX	Infancy	162
X	School Days	188
XI	Growing Up	207
XII	Courtship and Mating	224
XIII	The Birth of a Family	237
XIV	Social Poverty	247
XV	Social Sufficiency	275
XVI	A Community Grows	289
Appendices		299
Index		325

To those in despair of Man's civilization

A PARABLE

A lawn, a rabbit hutch, a much loved rabbit hopping about free in the sun. Its owner, a little girl, has heard a noise that fills her with dismay. She rushes out to find that the terrier from next door has escaped into her garden . . . loud barkings, a horrifying scuffle. The inevitable is happening . . . she flings herself to the ground, for she cannot see that dreadful end.

Minutes pass, blackness, abysmal horror, when faintly a voice reaches her, "It's all right, Jennifer, the rabbit's safe". The child uncovers her face; slowly she approaches the hutch: no cry of joy; she turns away in contemplation. Five minutes later she is heard saying to herself . . .

"I must remember, *always have a good look before you cry*"

HISTORICAL

Before beginning to build, it is necessary to know what bricks are to be used, or, in modern terms, what must be the *unit* of construction. Times and fashions change and with them the units of material construction. So, too, with the constructs of Society; man changes his institutions, his customs and the external circumstances of his life and, in a manner, his own life with them. But Nature's laws are abiding. In the realms of Matter and Energy about which man has come to know so much, he accepts Nature's units of construction and works in obedience to her laws. In the realm of Living he has yet to recognise the unit with which Nature works; and to learn to use that unit. If man is to venture on the rebuilding of Society, he must take nothing for granted. The first question therefore is—With what *unit* does Nature build in the living world?

It is with the answer to this question that this book is concerned, and because that is its subject we publish now what would in less turbulent times have been withheld until the studies that underlie it had reached a fuller measure of maturity. We claim to have defined the unit of Living. It is not the individual; it is *the family*. This has opened up a new field for experiment into social organisation and has enabled us to contrive the first rude instrument for the exploration of new possibilities. It is not then on theoretical grounds alone but with some basis of experiment and experience that we offer to the student concerned with the structure of Society of the future, an indication of what living unit will give us a living human Society—no matter what variations in detail may be necessary to suit different peoples and different climes.

In Science one thing leads to another. When setting out on a journey of exploration no scientist knows what his ultimate destination may be. Studies on the ill-defined fringes of several branches of Science, particularly that of pathology, have led to new light being thrown, not on Sickness, but on Health. The first indicating finger pointing to the necessity for the study of *health* came from work on the epidemiology of infectious disorder.

The unfailing resistance to infection of a proportion of all experimental animals in colonies—or individuals in a community for that matter—*unless previously subjected to injury*, demonstrated that conditions need to be extremely inimical to living before animals acquire a susceptibility great enough to yield a high incidence of infection. Susceptibility then was not 'normal'; it was an acquired characteristic. Following this line of reasoning, it became apparent that the process of Health was itself a subject worthy of study—one distinct and different from that of Sickness. The Science of Health—or Living—and the Science of Pathology are, as it were, the head and the tail stamped on the same coin; operating upon the same substance the pattern of the respective processes is different: they are antithetic sciences.

Another approach to the study of Health as opposed to that of Sickness and its causative agents, was derived from observation of the varying degrees of resilience of the human individual in the face of defect and disorder. Serious and even mortal disorder may be present and yet the individual, far from becoming a 'patient', may remain ignorant of anything amiss.

It must be remembered that in the hospital, which forms the main field for the study of sickness, only advanced stages of *incapacitating* disorder—that is to say, disorder of which the patient is himself aware—are available for study, while the main source of material for the pathologist has always been the post-mortem room where the causes of *death* rather than the causes of disorder are the most obvious findings. So, the student of the causation of disorder is persistently faced with the practical difficulty of being unable to come into contact with its early stages; that is to say, with disorder at the point of its departure from health. He is, hence, unable to study the change-over from one process to the other. So it became necessary for the purposes of study of the causation of disorder to create a field in which the presumed healthy could be observed, with a view to disclosing insidious disorder. It was thus largely the logical pursuit of the study of the causation of Sickness that presented the opportunity for the study of Health itself.

Lastly, certain considerations arising out of the development of Hygiene and Public Health began to compel adoption of more rational practical measures. The palpable irrationality of seeking Health through the current methods of Child Welfare affords an instance. In existing circumstances the consecutive care of the infant only begins *after* its birth, whereas it has been growing from the moment of conception. Neither is there any provision whereby the advantages of modern medical science can become available to all potential and prospective parents *before* conception, so ensuring that the child shall inherit its birthright of health. The hygienist's science, and his art which is essentially that of education, make it necessary to find a means of coming into contact with the family before parenthood ensues.

Thus, led on the one hand by the many straws blowing in the winnowing of pathology, urged on the other hand by practical considerations concerning the application of modern scientific knowledge for the development of health, a new position has been established and a new technique evolved.

How did it begin ? Very informally, very modestly. A small group of lay people, all under 30, had what might be called 'a hunch' that health was the factor of primary importance for human living. Like everyone else, they had only the vaguest notion of what they meant by 'health', but sensed that its secret lay with the infant and its early development. They were convinced that it mattered that parents should be free from sickness before the child was conceived and carried ; certain that the parents should *want* the child, and that they should be able and eager to rear it.

That sounds very commonplace. It was the action taken by this group that was important. They sought the advice and guidance of a scientific staff with a background such as we have described. A line of procedure was determined upon. It was decided to offer to families a *health* service constituted on the pattern of a Family Club, with periodic health overhaul for all its members and with various ancillary services for infants, children and parents alike.

The first question asked was a simple one. Were there families who, if it were offered them, would use a health service to enable them to keep fit and ward off sickness *before* they were smitten with incapacity ? It was argued by the sceptical that since the patient, even when ill, would often not go to the doctor, it was unlikely that families unaware of disability would willingly embrace periodic health overhaul. Let it be put to the test.

So in 1926 the pioneer "Health Centre" took shape.[1] A small house was taken in a South London borough. It was equipped with a consulting room, receptionist's office, bath and changing room, and one small club-room. Families living in the vicinity were invited to join this Family Club for a small weekly subscription. By the end of three years, 112 families, i.e., some 400 individuals, had joined and all the individuals of these families had presented themselves for periodic health overhaul. Not all had retained their membership throughout that period, but the question had been answered. *Given suitable circumstances*, there were families who would welcome a Health Service distinct from any sickness service and without being urged by any sense of impending sickness.

But perhaps the most outstanding fact learnt by the scientific staff was that although sickness could be detected early, often indeed long before the individual had any idea of its presence, and although it was found that the individuals subject to such disorders were willing and anxious to have them removed and with the assistance made available now took the necessary steps for their removal, it was in many cases useless to eradicate the disorder only to return the individual to the environmental conditions which had induced it. Equally important, it was discovered that in those who manifested no disorder, the standard of health or vitality found was low and could not be raised without suitable equipment for the purpose. In other words, it became clear that while operating efficiently as a sieve for the detection of disease and disorder, *periodic health overhaul is ineffective as a health measure in the absence of 'instruments of health'* providing conditions in and through which the biological potentiality of the family can find expression.

This finding was unforeseen. The issue now became greatly complicated. The possibilities both for the study and for the cultivation of health were opening out and taking on a new aspect. It was decided to shut down the first small Health Centre which was, as it were, a bench test, and to devise an experiment in which not only the technical measure of periodic health overhaul could be employed on a larger scale, but in which there would be available circumstances and material likely to kindle the health of the families examined.

[1] The Pioneer Health Centre, Queen's Road, Peckham, London, S.E.

Seven long years passed. They were spent in planning the next stage in great detail and in collecting money for a new and larger enterprise—a field experiment it might be called. It was to be a Health Centre to cater for 2,000 families, in which were to be offered consultative services as before, and in which the member-families would find equipment for the exercise of capacities for which there was little or no possible outlet in the ordinary circumstances of their lives.[1]

Thus in 1935 the second stage of the Pioneer Health Centre took form. It was a great venture : a social structure to be built with a new unit—not the individual but *the family*.

After eighteen months' work an interim report was published under the title of *Biologists in Search of Material.*[2] In that book many details of the initial procedure, including the technique followed in the periodic health overhaul, were given. These we do not propose to repeat here, for many will have read them, while for those who have not, the report is still available.

We must, however, emphasise the major conclusion of that report. It was that the Health Centre with its peculiar technique for dealing with families in a social milieu with simultaneous use of the periodic health overhaul, had provided us with an instrument of analysis not unlike a prism, which interposed in a beam of white light analyses it into its component parts giving a picture of the spectrum, or rainbow. The spectrum that the Centre revealed was, alas, no gay coloured rainbow. It made clear that the populace was composed of three categories[3] :—

1 those in whom disorder was accompanied by dis-*ease* 32%

2 those in whom disorder was masked by compensation, and who therefore appeared in a state of '*well-being*' 59%

3 those in whom neither disease, disorder, nor disability were detected—? the 'healthy' 9%

The mechanism of the Pioneer Health Centre has made it possible to view and to study discretionately, in the light of these several categories, so-called 'normal' families and individuals going about their daily business.

[1] For an account of the first Health Centre, and of the sketch plans for the second, see *The Case for Action*, Pearse & Williamson. (Faber & Faber, 1931). 2s.

[2] Faber & Faber, 1938. 2s.

[3] *Biologists in Search of Material*, p. 78.

As the experiment has proceeded, the understanding of the scientific staff has deepened, and the theory of Health has been developed and clarified. We can now visualise the essential elements of a technique for the *practice* of Health as something different and distinct from the practice of Medicine. It is with this subject that the greater part of this book is concerned.

The first three chapters range over a wide field, affording a sketch map of the territory into which we are being led, and indicate the principles of growth and development that are beginning to stand out as fundamental to Living. To some, these chapters may seem difficult, leading them into realms with which they are little familiar. Those readers may prefer to skip this portion of the book and pass on to what for them may seem a more human aspect of the experiment, only returning later to the theory—as one returns to the map at the end of a day's journey—to trace the path along which they have travelled through a close-woven chronicle of human circumstance.

So vivid was the life, so illuminating the understanding that came to those who worked and moved in and with the experiment, that it remains unobliterable. The war, passing like the black shadow of an eclipse across the world, has caused the experiment to be suspended,[1] but live and vibrant beneath what is now a scorched earth, "the Centre" lives to thrust up in a new age.

It has already proved itself a 'living structure'.

[1] The Centre's activities were suspended at the outbreak of war, September, 1939, owing to the inevitable dispersion of the family unit in war conditions.

CHAPTER I

LIVING THINGS

Living cannot be interpreted in terms of materio-dynamics. Some other cosmic principle is at work not included in and not defined by the laws of matter or of radiation, however "deep the waters" into which the study of these have led. "The universe can be best pictured, although still very imperfectly and inadequately, as consisting of pure thought, the thought of what, for want of a wider word, we must describe as a mathematical thinker".[1] Possible though it is to conceive of thought without personal attributes, that concept is nevertheless impossible divorced from the quality of livingness. But this picture does not disclose to us the nature of the livingness behind thought; it only implies that when the laws of Living are disclosed they may demand yet one more co-ordinate on our graphs and yet another mathematic even more intricate than the last.

Setting out as biologists to study Living, we shall not attempt to define the nature of Life any more than the physical scientist defines the ultimate of Energy, but merely proceed to examine its manifestations in the living entity and to determine the laws that underlie its operation.

But for the manifest of Life we have as yet no exact name, that is to say, we are not any more discriminately aware of its nature and identity than man must always have been of the sun's rays before Science began to work its own magic with them. So, before proceeding to grope our way forward in the study of the manifest of Life we must name it. Throughout this book we propose to call it '*function*'.

Appropriation of the word 'function' for the manifest of Life raises with workers in other fields of Science an issue demanding clarification. The physiologist, long first in the field in the study of the mechanism of the living body, uses the word 'function' freely, and in our understanding indiscriminately, to cover several distinguishable operations of organism not all of which are significant in the study of living. To illustrate this we might refer to one of the common technical procedures of the physiologist

[1] *The Mysterious Universe*, Sir James Jeans. (Cambridge University Press, 1930), p. 136.

where, by excision or other means, he isolates organs from the general influences of the body in which they occur, and *fixing their environmental conditions*, proceeds to observe their 'functional' response under given stimuli. Perhaps the best known and extreme example of this type of procedure is the excised chicken-heart which was kept beating in a bottle some twenty-five years after the death of the chicken. This heart ceased to beat and 'died' owing to a single oversight in which there was failure to adjust the perfusing fluid,—evidence that its 'survival' was dependent upon the rigid fixation of its environment.

This brilliant experiment gives us information about the mechanism of heart muscle, but it gives none either of the *functioning* heart or of the living chicken. It is not unlike the information gained from the bench test of an internal combustion engine. A bench test is a valuable test, but it gives no indication of the final performance, for example, of an aeroplane in flight,—a tool in the hands of a skilled pilot instant to adjust the machine to the least suspicion of environmental change. The conditions imposed upon the engine by the will of the living pilot in response to changes he encounters in the environment will to a very large extent nullify the value of the information gained from the bench test for what is commonly called 'practical' use—for which the engine was in fact invented. Certainly the inferences to be drawn from the performance of the engine in these two circumstances are not identical. With most physiological studies this is no less true ; they yield information about the body but not about its 'living'.

So, two distinct and different studies may be made : one of the response of organism, organs or tissues in a controlled, artificial or fixed environment—*physiological operation* ; the other of the behaviour of the living organism as a unity in an ever changing and free environment—*biological function*. It is for the latter that we shall consistently reserve the word 'function' in this book.

The next step is to determine through what unit function is manifest ; and where and how it can most easily be studied.

A unit is the smallest 'parcel', aggregate or organisation which exhibits the characteristic attributes of any substance, potency or entity. Technically, living entities are called 'organisms'.

Before, therefore, we can answer these two questions, a further question must be asked: What is an organism?

By 'organism' we understand any living entity capable of performing the full cycle of its specific existence. Not all living entities fall within this definition. For example, a soldier-ant is a living entity but it is not an organism, for alone it is unable to complete the cycle of ant-hood. In the ant species various operations integral to ant are delegated to various entities in the heap. The queen alone can lay eggs; the soldiers protect the queen; the workers feed her, etc. Each entity has its own special work to contribute to the organism, and without that contribution the function of ant is abrogated and continued life in the organism, 'ant', ceases.[1]

The ant-*heap* alone represents the full range of function of ant-hood. It is then the ant-heap that represents the unit-organism 'ant'. Similarly, it is the hive of bees—not the single bee representing a specific operation essential to the hive, or colony—that forms the organism 'bee'.

"And for their monarch Queen—an egg casting machine
Helpless without attendance as a farmer's drill,
By bedels driven and gear'd and in furrows steer'd
Well watched the while, and treated with respect and care
So long as she run well, oil'd stoked and kept in trim".[2]

Bridges knew the Queen to be but the *ovary*—a mere organ of the organism—'bee'. But here in the hive of bees there is so extreme a degree of separation of the respective organs of the whole organism that the casual observer has been deceived into regarding each bee as a separate organism.

Seen from the same aspect of function, two frogs, male and female, compose the organism 'frog', for though one frog or one soldier-ant is an integral part of the organism 'frog' or 'ant', neither alone represents the unit-organism capable of the full functional cycle of their species. This is so obvious that it may be wondered why we stress the point. It is, however, of the greatest importance to the student of function, for were we to study male frogs or soldier-ants alone, in ignorance of their connection with the facultative species 'frog' or with the ant-heap, we should never arrive at a knowledge of the full functional

[1] See *The Soul of a White Ant*, Marais. (Methuen).

[2] Bridges. *The Testament of Beauty*. (Clarenden Press, 1929), p.55.

capacity of their respective species. The part cannot declare the function of the whole.

As students of *function* in man (*homo sapiens*) we must then at the outset be careful not to take anything for granted: not to mistake the individual for the whole organism, for, as we have seen, the individual may be but an organ of a more complex organism. By a mistake of this order we should be doomed to miss the manifestations of function that we are seeking.

In studying the mechanics and dynamics of the human body this point is not of the same critical significance. For instance, the student has merely to make an adjustment for the sex of the individual studied, to arrive at a knowledge of the mechanism of the body of the species. If he is conversant with the anatomy of a man it will serve—with the addition of facts about the difference in weight, shape, structure of bone, etc., together with a knowledge of the difference in the sex organs—for a knowledge of the anatomy of the human species. What knowledge he has of the bio-chemistry of the alimentary system can be applied with success to either sex indiscriminately. When we come, however, to function, this method no longer serves, for we find that man and woman are not functionally identical entities exhibiting merely superficial differences aligning them for co-operation, as in the reciprocity of mechanism; not merely two entities with capacities so nearly equivalent that they can shoulder the same tasks and by means of a statistical correction be regarded as interchangeable units, as, for example, in the 'science' of economics, in industry, or in the labour market. In the functional sphere man and woman do not work reciprocally as in mechanism, but *mutually* as diverse parts or organs of a unified organism —like a small ant-heap linked in the continuity—or what later we shall have to call the 'specificity'—of a 'functional organisation'.

After mating has occurred, invoking a new functional organisation, we no longer have a man and woman who, shackled like the links of a chain, have joined hands in marriage, but one bi-polar unity—with maleness and femaleness at its opposite poles. How can we visualise such a unity? In the physical realm it is perhaps not unlike the solution of metal within metal such as we find in some of the amalgams. Or, in the physiological realm, perhaps we are led to recall the bi-axial construction of the features of the human body: right and left handed, right and left kidneyed, right and left hearted, right and left eyed. The unity of the mated pair, dual like the body, is right and left individualled, as it were. Thus, the human organism, like the body itself, is a *unity* balanced in function as in feature.

The reader may perhaps find this a difficult conception to grasp; may object that any process of merging of two individualities suggests loss rather than gain, and hence is one that cannot represent the true picture of progressive human functioning. On the contrary, the new polarity of the functioning organism brings with it for each individual a measure of fulfilment unobtainable by either alone.

We know the opprobrium implied in the expression 'one-eyedness'. This is not without reason. If we look more closely at the mechanism of optics we see that each eye looking separately sees a field more limited than that covered by the two together. But this is not all. Binocular vision does not merely *reproduce* in a combined and enlarged picture the field of view of each separate eye. The two eyes acting as a unity create a novel image. So there emerges the 'solid reality' of a stereograph, which no one-eyed vision can achieve. What applies to vision seems to apply to all functional action: it is dependent upon duality operating in unity. So too with the mated pair we find duality operating in the unity of male and female. Hence man also is bi-polar in function. There is no sacrifice here; neither is it compromise. Just as the eyes in binocular vision produce a stereograph, an origination or novelty, so it is the 'parenthood' engendered by the unity of two diversities—mature manhood and womanhood—which originates, or brings the *new* to birth.

Thus when two diverse individuals function as an organism, *all that they encounter* acquires a new significance. It is not merely the addition of the experience of one to that of the other, making the combined view a larger whole seen, but that with new polarity a new *quality* is given to their apprehension. And this quality of perception is given not only to what is experienced at the moment, but that experience itself influences what they in their new functional orientation will in the future experience—hence altering their every action.

The supreme and most concrete example of such an origination or novelty from the fusion of two diversities is, of course, the child. It is a reproduction neither of mother nor of father;

indeed, not a *re*production at all. It is a new and unique individuality that is originated through the bipolarity of organismal function.

So it is through the unified mutual action of two entities, man and woman, that alone the full function of Man is manifest: that full and rich diversification of his species proceeds, and that human potentiality finds its full expression. Thus while the individual man (or woman) is a satisfactory subject for study by the zoologist, physiologist or pathologist, only man-and-women as a unity can meet the needs of the biologist setting out to study function.[1] What then are we to call this functional unity—this concept of the biological unit? We have named it 'family' implying by that word the mated pair either with or as yet without children, and it is in this sense that the word '*family*' will be used throughout this book.

There are other difficulties and subtle snares with which biological material confronts the student. Function is not always explicit. Like force, it may be potential or latent; that is to say, the full range of functional manifestation of the species may only become explicit in certain circumstances the nature of which we do not yet understand, and which may well chance to be absent when and as we observe. We may, in fact, only be familiar with rudimentary manifestations of function in the life of any species. In ignorance of any fuller manifestation, how easy to take these to be the full expression of its potentialities. A striking example of such a situation is to be found in the case of the Mexican axylotl, tadpole of the salamander (*Amblystoma*). This large aquatic tadpole can and usually does live, breed, rear its kind and die in its unmetamorphosed (tadpole) state. Only many years after it had become known to the zoologist and familiar as a fashionable parlour pet was it discovered to be merely the tadpole or larval state of a land-walking salamander catalogued as a different species. Because the axylotl was able to live and propagate its kind in its immature form, its potentiality for living a different and wider existence, for acquiring lungs and walking on dry land, was missed even by the zoologist.[2]

[1] See also *The Case for Action*, p. 60-65.

[2] It is possible in the laboratory to effect the metamorphosis from tadpole to salamander within a few days by the injection of thyroid extract. From this we must infer that the potency of endocrine secretions cannot be over estimated in their effect upon function.

To snares of this nature the experimenter must be alive as he approaches the field of function in human biology, for he may not assume that man as we now know him is man whose potentialities have already found their full expression.

We have already shown in an earlier publication[1] that there are three distinct states in which man may exist while carrying on his daily life. Hampered by disorders he may suffer from the ravages of disease; cloaking his disorders by the use of his reserves, he may be buoyed up by a false sense of 'wellbeing', or, lastly, he may live a full functional existence in which his development is proceeding according to his potentiality. The difference between these three states has been shown to depend on the several relationships of the individual to his environment. In the last of these three states only, is man free to act in mutual response to an ever changing environment. It is this last state which we recognise as the legitimate field of *Health* or 'wholeness'. This field of Health or sanity with which we as biologists are concerned will be found to be distinct from that of Sickness, where subject to disorder the individual obeys the laws of pathology. In Health man observes a different natural law:—the law of *function* with which we are here concerned.

It might with reason be asked:—Why entertain the idea of using man, with all his complexity, as the experimental animal in what is so new and difficult a field, for surely the first necessity is to find the simplest organism for investigation? The zoologist hitherto concerned with the classification of species and the particulate description of living entities, has turned naturally to unicellular entities having little anatomical structure other than a delimitating membrane enclosing a nucleus surrounded by a body of cytoplasm, such for example as amoeba, or paramecium.[2] These are the simplest for his purpose. But when attention is turned to function the scene begins to shift. When the amoeba encounters food in the immediate environment the *whole* entity flows towards the attractive morsel; it stretches out its body in the form of embracing limbs—pseudopodia, surrounds the food particle, and, dragging its whole body forward in the direction of its embrace, engulfs the prize. Whatever attracts

[1] *Biologists in Search of Material*, p. 78 et seq.

[2] For a popular account of amoeba see *Cine-Biology*, Durden, Field & Percy Smith. (Penguin 1941).

it, the appearances to all intents and purposes are identical—an all-or-nothing type of enveloping action for each and every new experience embraced. How confusing to the observer this apparent similarity of expression for all the delights of life!

In order to observe functional action in its discretionate form we are forced to the opposite end of the zoological scale and it is to Man himself we turn. Man wishing to eat can take his food with finger and thumb and while doing so can carry on simultaneously many other distinct and intricate operations. Five fingers have been differentiated in his hand, each capable of separate co-ordination to effect discretionate movement; he has acquired a constant and material gullet, stomach, liver; he has a renal system and complete and well-defined nervous system, etc.—all of which have acquired through age long differentiation of his species a high degree of special and independent action. The human organism then, is the most convenient primer for the biologist who as student of function seeks to elucidate the laws of living.

What of the next question? How does the biological organism or any lesser biological entity proceed to its fulfilment through function? We have been accustomed to regard the living entity at its inception and in its most primitive state, as no more than a focus of livingness in a limitless and apparently passive and wholly *unfamiliar* environment. There is interposed between the two—that is between the entity and its environment—no more than the semblance of a membrane created by the difference in direction and rate of two dissimilar motions. Thus we usually visualise the simple cell, amoeba. Thus we marvel at the sureness with which the unprotected speck of protoplasm, presumptively unguided and born into what is usually conceived as a 'hostile' environment, shapes and forms itself with such unfailing accuracy and, acquiring the specific features of its kind, reaches maturity. Whence comes the material for its growth and development? It is from its environment that has been presumed to be hostile.

The picture of the amoeba lured to engulf a particle of food in its nearby surrounding medium is diagrammatic of the process of accretion in all living entities. The amoeba embracing a particle from the environment, engulfs the morsel and digests it. On such meals it lives and grows. This tells us the source of its increment: all material for increase comes from the environment. It tells us nothing, however, of the method by which the individual converts the ingested environmental moiety into the substance of its own body. So, before the significance of the above picture can be understood, the fate of the engulfed environmental contribution must be followed.

Once within the body, the morsel is picked to pieces, chemically analysed, sorted out and separated. Certain selected portions are then as it were reshaped and woven into its very substance according to its specific order, thereby adding to and developing its unique basic design. This process—the living power to build up a basic organic design from the substance of the environment—is called 'synthesis'. The process of acquisition is the same whether it be of food, light, or any other engulfed 'experience'. Once of the body, all is stamped with the trademark of the receiving house, part transitorily, part, and that a highly selected portion, indelibly marked or 'sensitised' with the individuality of its new host.

From this 'factory' there is an enormous output, some to be sure consisting of rejected intake; some the product of physiological work done—heat produced, etc. The higher we rise in the zoological scale the more important and distinctive becomes the other surplus of highly elaborate biological synthesis, for owing to the specificity it acquires in the 'factory' through which it has been processed, it is potent in the environment to which it is returned—leaves, fruits, hoof, skin, hair, urine, faeces, etc. In his intuitive wisdom in the past, the good cultivator of the soil has for centuries known the value of these so-called 'waste' products for maintaining fertility. It is modern civilisation that has applied the word 'waste' to them. To the physiologist also they are 'waste'—a nuisance—for their excretion brings about changes in the environment which he is seeking to stabilise and fix for the conduct of his experiment. Perhaps indeed it was the physiologist's original use of this word 'waste' for goods returned from the body's factory, that has been responsible for hindering our appreciation of the significance for *living* of these materials returned to the environment, and hence of their significance to the future of the living organism itself.

In the field of function where individual and environment work in strict mutuality they assume an importance of a magnitude not yet recognised. For all the products of work done must be included in this category—the objective products of the living entity's subjective process of development. When we come to Man his surplus of synthetic products cast into the environment is well nigh overwhelming, for, besides his physical excreta, there are the products of his mind and of his skill: his inventions, his music, his art, his science—all shed fruits of his synthesis. Let an apple fall, and as a result a Newton sets in motion a train of further human synthesis that changes the face of man's world!

Thus, the environment is the source of diversity as well as the recipient of the diversification of that which is taken from it by the organism. Each different factor or change in the environment that impinges on the organism, each new food particle digested, each new co-ordination learned as a result of experience made possible by any new environmental disposition, results in the development of further specificity in the organism and leads to a still more versatile power of apprehension of further environmental contributions. Also, and consequently, it leads to still further novelty in the products subsequently received into the environment. So that in the presence of adequate nutriment, function implies an ever increasing diversification, in the organism and in the environment alike.

This is the functional picture of life in flow. It is to be seen in a progressive *mutual synthesis participated in by both organism and environment*. It is wholeness—Health.

Here then is a picture not of hostility between the organism and its environment but of *mutuality* at work in the living world. Yet hitherto the development and indeed the very existence of the organism has been pictured by the scientist as a 'struggle' for survival[1]; while Man in his acknowledged supremacy is alleged to have 'conquered' Nature, i.e. his environment, rather than to have wooed her in the sensitivity of a mutual—or loving—relationship.

It is in the modern revival of ancient methods of agriculture[2] and in the science of oecology that hitherto there has been the fullest appreciation of the mutuality of function in the organism,

[1] Cf for example *Man on his Nature*, Sir Charles Sherrington, Chapter XII, *Conflict with Nature*, p. 359. (Cambridge University Press, 1940).
[2] Cf *The Living Soil*. E. B. Balfour (Faber & Faber, 1943).

with its essential shuttle-like throw from environment to organism and from organism to environment, each throw changing the design, each change affording a stimulus to the next change that is to follow. Yet plant, animal and man live by the same biological law. The laws that govern growth and development apply equally to the organism as a whole, or to its parts.

As we have proceeded, there may have been gathering in the mind of the reader a growing speculation, perhaps an almost nervous apprehension, as to the constitution of the environment itself in this mutual transaction. Neither has this thought escaped the biologist. He begins to appreciate that the process of diversification so characteristic of organism and, as a result of the life process, equally apparent in the environment, must denote some *progressive order* in the latter. Can it be that the environment, also 'in process', is taking on an orientation as ordered as that which the embryologist can follow so clearly in the differentiation of the embryo—like the chick developing from the amorphous material of the egg? Is, then, the process we call 'evolution', with all its manifest expressions, but one universal expression of the 'organ-ation' of the environment itself? Is the environment *alive*?

The mutual action of organism and environment, associated as we rise in the biological scale with an increasing degree of autonomy of the organism, recalls forcibly to mind the circumstances of a single cell, such for instance as the liver cell, set in the body of which it is an infinitesimal part. The cell acts as liver cell carrying on the specific function of 'liverness', yet always, in health, 'aware' of, and subject to, the wider needs of the body of which it is part and from which it derives sustenance. It is this *relationship to the body* which alone gives significance to its individuality as liver cell as well as to its unique function of liverness.

The pathologist is only too familiar with the situation that arises where this delicately poised relationship of the cell's autonomy within the sphere of a greater organisation—the body—is absent. When the cell multiplies without reference to the impulses of the greater organisation of the body of its inhabitation, the result is cancer, the definition of which might be stated as 'multiplication without function'—loss of individuality. Such procedure ushers in antagonism, disrupting the mutual association between the cell and its environment—and ends in the ultimate destruction of the cell, of the body in which it grows, or of both.

Thus the body as an organisation is, in fact, the ultimate significance of the cell. Can it then be that Man himself is but a cell in the body of Cosmos ; and that Cosmos is organismal as he is ?

Without being able to define the factual basis for their intuition —for that can only come through science—wise men in all ages have acted with a deep intuitive consciousness of this as a truth. Upon it they have built their hopes, their conduct and their religions. Only now, as intuitive apprehension seems to be wearing thin and threadbare, are men of science being led, through the study of function, to suspect that there may even be a physical basis for these primitive intuitive actions ; that in fact the significance of human living lies in the degree of *mutuality* established with an all pervading order, Nature—whether we deify her or not.

So in order to study function, we must turn to the organism and its environment, in process of mutual synthesis ; the organism of choice for our study being the human family, *homo sapiens*. Herein then lies a challenge to adventure. The student at this point must be willing to put from him the comfortable cloister of the traditional laboratory where he has learnt the structure and the classification of species ; he must leave the protected harbours of the physiologist where the merest zephyrs of the environment are steadied and controlled ; he must part company too with the student of medicine who, informed by the science of suffering (pathology), has been searching for negative evidence of function (health) in the shadow of sickness and amidst the shades of the dying. Leaving these behind him, now as biologist prepared to sail upon the open sea of humanity where the manifold winds of the environment play in ceaseless change, he may set out on a further search into the science of Living.

CHAPTER II

MAN IN THE MAKING

We have seen the child arising as an origination, a 'novelty', out of the functional re-orientation of two individuals into a family; arising not by a process of *re*production as the analytic and classificatory phase of biology has represented it, but as a true creation. Indeed the man in the street has been wiser than the scientist, for he has always said that genius[1] is "born not made". Here he has grasped a truth, for the essence of birth, unlike the repetition of mechanism, is *individuality*—that uniqueness of the living entity which is the hall mark of function.

How little concern do we give to this great and outstanding characteristic of *livingness*. How little, for instance, do we account it as strange that we and everyone else can recognise as distinct the fifty Mr. Smiths who live in our town, or that no two individuals have identical finger prints. Stranger still, the dog can tell not merely the sound of the approach of his master's car, but can recognise that it is his master who is driving that car. Not only are a man's features individual, but so all pervading is his individuality that it gives uniqueness even to the pattern of his action which, like the pebble in the pond, according to its quality sends forth its ripple into the environment, or pool of cosmos.

Surely here is a mystery, something to be enquired into: the functional behaviour of the individual demonstrating uniqueness, the signature of which is so indelibly imprinted both upon his anatomy and upon his action-pattern. What then, we must ask, is the craft by which man fashions his individuality; what the method by which Nature achieves this marvel of diversification in cosmos?

Not by studying the machine and its mechanism—even man's own machine, his body—can the secrets of birth with its power to generate novelty be disclosed. While in the chemico-physical realm analysis is the way to understanding, in function it is through synthesis. In other words to see how a machine works you take it to pieces; but to see how a living entity functions it

[1] Genius—"The tutelary god or attendant spirit allotted to every person at his birth to preside over his destiny in life". Shorter Oxford Dictionary.

must be seen in its organismal unity and in its living environment.[1]

It is then not to the individual but to the *family* as we have defined it, that we must turn our attention in order to study function. Only in the study of family can the ' how ' of living process show itself ; can the ' how ' of birth and differentiation with the production of infinite diversity be declared.

If we survey man's actions and behaviour we see that they are directed by two types of wisdom ; voluntary wisdom and involuntary or *autonomic* wisdom. It is this latter which for example guides the heart beat, the respiration, the processes of alimentary digestion and takes over control of the body in sleep. In the use of his voluntary wisdom, man is as it were ' given his head ' to find his own way and effect a synthesis with the maze of new circumstances that continually arise before him. But in the use of his involuntary wisdom he treads the great highway of organismal experience established by the use of countless generations of living entities before him and sealed by Nature as valid for the furtherance of function. Thus if we wish to find out *how* Nature works in the field of human function, we must first examine functional action guided by involuntary wisdom unconfused by man's voluntary control.

Perhaps the clearest picture Nature offers of the procedure of function under the guidance of the autonomic wisdom of the organism, is to be found in the development of the embryo in its maternal environment. Let us then turn our attention to the ovum and follow its course of development through pregnancy.

The ovum arises from the cells of the ovary. It first becomes recognisable as an individuality as it lies free in the fluid of an ovarian follicle which forms about it as it separates off from the surrounding ovarian cells. We know that the tissue-fluids of the ovary have qualities of preponderating feminine bias secreted from the woman's blood by the glandular mechanism of the ovary, for extracts of this tissue yield powerful and specific effects when administered by the mouth or by other routes. It

[1] " We cannot possibly examine separately the parts involved in life as we examine separately the parts of a machine. In particular we cannot separate the influence of the environment since environment belongs to the unity which we perceive as life".
The Sciences and Philosophy, J. S. Haldane. (Hodder & Stoughton, 1927-28), p. 81-82.

is in the influence of these fluids that at its outset the ovum lies bathed. Here it is reared until it is ready for adventure, preparing itself, by a process called 'maturation' for its coming union with the sperm.

In its nucleus lie the chromosomes carrying the genes that are to play so important a part in the transference of certain characteristics to the new individual. These chromosomes of its nucleus divide, half their number only remaining in readiness for the completion, qualification and enrichment that is to come through fertilisation. The mechanism of this division and reassemblage in preparation for the male contribution is well known, much work having been done on the subject.

But what of the other half of the chromosomes ? Together with a small part of the cell protoplasm of the ovum they are cast off in the form of what are called 'polar bodies'. This chromosome moiety, the other half of which contains the genes of such peculiar importance to the embryo, is returned to and absorbed by the maternal body. Since these polar bodies disappear, they have until recently been considered to be of little importance in development. In view, however, of the general principle of mutual action as a characteristic of function, it is possible that the polar bodies are in fact in the nature of a potent (endocrine) secretion of the ovum. It is at any rate clear that the ovum's *ex*cretion has become the mother's *in*cretion. The ovum has in fact left a moiety of very important substance—a bit of itself—in the ovary as a memento of its origin, before setting out on its journey of development. Thus, as it proceeds with its *own* individualisation it has a means of informing the mother of its achievement and so of maintaining the functional organisation of their unity.

Already now an individual, this ovum, like its mother, has its own secretory process. So it is no surprise to find evidence within the ovum itself of the concurrent accumulation of an *in*cretion, upon which it is to draw for sustenance during the next stage of its journey from the ovary to its nidus in the womb. We are very familiar with this incretion in the egg of birds, for it is represented by the yolk, or yellow, contributed to the egg by the ovum itself.

Wholly .untutored and inexperienced though it is usually supposed to be, the ovum thus does not set out on its journey either unprepared or unprovided for. This yolk stored away beside it—like the dowry of the bride taken from the old home to form the basis of a new one—is to serve for building its first born cells in the womb. The ovum has provided itself with a 'knapsack' filled from the home larder, which is to serve for its period of *initiation into a new environment.* The ration provided varies in bulk according to the length and nature of the next stage of the journey. In the human ovum the stage between rupture of the ovarian follicle and nidation[1] is a relatively short one in which no great increase in size occurs, and the material provided for the journey is a small one—small, but still indispensable to development.[2]

We can better appreciate the significance of this provision in some of the lower species where the ovum, early cast out of the maternal body into its new environment, is destined to lead a more independent existence. In the hen's egg, for example, the relative sizes of the embryo and the knapsack are reversed, for here the embryo is no more than a small streak lying on the back of its comparatively huge knapsack—the yolk of the egg.

In the hen's egg we see that a further interesting thing happens. As the ovum passes on its journey from ovary to nest, the mother hen (from the excretions of the oviduct) also makes *her* contribution to its knapsack of food. This contribution is the white of the egg and the enveloping shell. The knapsack is now complete and the ovum, sealed up with its ration of food within the shell, is laid in the nest.

It is out of this ration provided by ovum and mother together that the shapeless ovum is to fashion its own body and emerge, an organised individual—in the case of the bird's egg so far as is known with the minimum of further help. Thus the embryo cries—"give me the *materials* and I will both fashion the tools

[1] Nidation: A technical term used for the implantation of the ovum in the wall of the womb: literally-'nesting'.

[2] It has not so far been possible to induce growth in any ovum robbed of its cytoplasm. The reverse, an ovum depleted of its chromosome elements, can be induced to subdivide to some extent. Both are necessary and must be attuned one to the other or development ultimately fails.

"The main thing to keep in mind is that *the cytoplasm of the egg does contain specific developmental properties.* Consequently, neither the nucleus alone nor the plasmatic cell body alone can be considered the primordium of a new organism; only *the egg as a whole* deserves this title".

Principles of Development, Paul Weiss. (Henry Holt & Company, New York, 1939), p. 185.

and finish the job—of producing a unique individuality!" It takes to the work like a duck to water, *lacking no essential wisdom from the earliest moment.*

At this point a question arises that has not to our knowledge been either asked or answered. Can it be that the most important factor about the provision in the knapsack is not so much the quantity of the food so cunningly provided for the traveller as the *quality* it represents? It is clear that this knapsack of provender is something of which the inexperienced ovum already has knowledge, something to which it has already been attuned or 'sensitised' in the maternal body. When the egg with its yolk prepared in the ovary, leaves the ovarian follicle and, acquiring white and shell in passage from the oviduct, reaches the nest, there is ready for the traveller a carry-over of *specific* nutriment from one environment to the next. And this is no isolated instance, but represents a principle at work throughout the animal and vegetable kingdoms. Even in the plant world we find a similar knapsack of food within the seed provided for the developing germ when it reaches the soil and begins to germinate.

It is as though the developing embryo takes with it and has at hand at each new stage a sampler of the work it effected in the previous one—a specimen of the digest of its last experience—to serve as a pattern or indication of the new processes to be evolved. So it goes forth on its journey of development, not wholly without guidance as is usually supposed, but informed by knowledge derived from its *nurture.*

We must now go back for a moment to the maternal ovary from which the ovum came. As a result of the departure of the ovum from the follicle a new ovarian secretion—'progesterone'—has begun to pour out from the vacated bed of the ruptured ovarian follicle. This floods the mother's blood, inducing fulfilment of her next cycle, and at the same time preparing a soft and succulent lining in the womb for the reception of the ovum as it arrives there from the Fallopian tube. So, stage by stage, mother and ovum work in complete mutuality.

But what now? Is all this careful preparation of the ovum for nidation and of the uterine membrane for the reception of the ovum to go forwards, or to go backwards? That hangs in the balance. Alone and isolated, woman and ovum can proceed no further. Some other factor, one of different polarity, must collaborate. The ovum must await its partner, the sperm.

While the woman is working in contiguous periods in maturing ova, the man works continuously in maturing the sperm, the secretion of which is stored for use as required, reabsorption occuring from this storehouse just as from any gland—a kidney, a liver or a thyroid. The sperm is thus kept as up-to-date as are the ova. The sperm has no visible yolk or food store ; what then is the endowment that it brings to the wedding ? Apart from its important chromosome content, we do not know. Perhaps its dowry is bio-physical (e.g. radiants of energy), rather than bio-chemical, as in the ovum—being as the spark is to the petrol in a motor. Certainly the sperm is motile and not passive, as is the ovum ; it is not swept along by external forces as is the ovum in its passage to the nuptial site, but is highly virile in its active motility. It appears and acts like a dynamic unit of the bio-physics of the body ; like the heat spark consumed or developed in an explosive process of growth, differentiation and development.

The male parent produces many potential bridegrooms for the ovum. While the single bride represents the specificity of its source of origin, each individual sperm no doubt has its own aspect of the specificity of its origin. Thus, though but one of the millions of spermatazoa is eclective to the bride, the others are not wasted ; all bring a gift to the wedding, for they are absorbed by the woman and we must suppose that their effect is as potent for her as is that of the chosen one for the ovum. All those unabsorbed by the ovum being absorbed by the mother must add the specificity of the male to the woman's biological and physiological economy. Thus, while the specificity of the ovum is directly met by the chosen sperm, absorption of the mass of sperm as an incretion by the woman renders them also environmental to the fertilised ovum and so part of the unity—foetus and mother—'the pregnancy'.

In the living world, Nature's usual method is to work, not by chance or luck, but by specificity. Thus the bride—the human ovum—is no doubt approached by the sperm as the male bee approaches the queen, i.e., by the exclusive electivity of its dynamic specificity. Autonomic wisdom in the bee has decreed a nuptial flight, and it may turn out that the human differs from the spectacular flight of the bee only in the closeness of the confines of its excursion. We do know that this 'nuptial flight' must take place within a certain critical period of the maturation of the ovum. The ovum is said to leave the follicle and descend into the tubes between the eighth and sixteenth or eighteenth day following cessation of the last period, and it is while in transit that fertilisation is thought to occur. In the present state of our knowledge, it seems therefore that fertilisation must be initiated by the woman's physiology on behalf of the ovum, her periodicity acting as switch to the process.

And so we come to the subject of wooing. This is not by any means the preserve of primary courtship. Wooing is no single incident in the life history of the pair but, so far as the electively monogamous are concerned, is a cyclical recurrent event. We see this clearly in many animals in the wild state and especially in birds, where the times for billing and cooing are well defined and coincide more precisely with the seasons than in the human species. The wooing process, by which the mutualising of specificities demanded by ova and sperm is achieved, is likely to be contributory to the progressive maturation of the parental unity.

But what happens to the ovum if no sperm arrives? Passing the zenith of its maturity it soon shows signs of ageing; its cytoplasmic substance decreasing in dispersion passes into a more consolidated or fixed form—a 'gel' phase—which toughens the egg, reducing its plasticity. Meanwhile, deturgescence of the lining of the womb is taking place; that is to say, it shrinks and consolidates again, shedding in the menstrual fluid the membrane which had been prepared to receive the ovum.

If, on the other hand, entry of the fertilising sperm does take place, the whole scene is instantly changed. Now the mutual process once more goes forward. In the substance of the ovum fertilisation brings about a phase of maximum dispersion—the 'sol' phase—which gives to it a high degree of fluidity and plasticity. As a result of fertilisation it reaches a state as long ago as 1876 figuratively described by Butschli as one of 'rejuvenation'—a view now meeting with confirmation in terms of coloidal physics.[1]

Unfortunately, we cannot follow by observation the course

[1] Cf Weiss, *opus cit*, p. 195.

of the human ovum from ovary to nidation in the womb. So accurately, however, can we follow the happening in ova of lower species, and so closely identical is the behaviour of all ova in the first stages after fertilisation, that we may assume that the human case is no exception.

With entry of the one elected sperm containing its own male chromosomes, coalescence of the male and female chromosomes again brings the number in the ovum to the correct number for its species. Now it is complete and capable of development. But at this point there is a pause in the proceedings when any observer sees but little ; quiesence ; no passage in or out of the cell ; no sign of its prospect. This phase is often described as a 'latent period' but we prefer to call it an inturning or 'centripetal' phase of activity. Because we do not as yet know what is going on, we cannot assume that nothing is taking place in the functional orientation of the new entity.

Following upon fertilisation of the ovum, there has been developing in the maternal ovary, from the bed vacated by the ovum, a conspicuous *corpus luteum*, or yellow body. This area of the ovary now continues to pour out 'progesterone' in greater abundance. So, step by step, the mother works mutually with the ovum to establish the pregnancy. If the maternal womb has not been carefully prepared, the unfortunate embryo cannot find a suitable site, so that even its implantation in the womb is a process involving mutuality of the embryo and its environment, the mother's body.

If the site is prepared, then quickly the ovum proceeds to establish a niche for itself in the acceptive and co-operative womb. Instantly there follows a burst of activity. Cells begin to divide, the whole structure to enlarge and soon an entirely new set of circumstances arises. From now onwards the embryo is promoted from existing on the contents of its first knapsack of accustomed food increted by the ovum from the tissue fluids derived from the maternal blood, to nourishing itself from the full flood of the blood of the mother. And here we must recall that, by the time fertilisation has taken place, the blood of the mother has been sensitised and attuned to that of her mate, father of the embryo. Thus, from the moment the new individual begins to develop, there is already a *physical* basis for a 'functional organisation' between the three individuals of the growing family.

So the developing embryo, building up its own unique individuality, is gradually led for its nutriment from its first familiar environment—that of the fluids of the ovary of its origin—to one of a wider familiarity—the blood of its mother—to which mother, father and embryo all alike have made their specific contribution. How careful, provident and gentle has been the nurture of this young ovum under Nature's tuition!

The next stage in development of the pregnancy is one of great significance for the student of function, for now we are to watch the functional organisation of mother and foetus converted for the time being into an actual anatomical unity, through the formation of the placenta or so-called 'afterbirth'. It is by means of the placenta that the two now proceed to carry on their synthesis in mutuality. Here then we are presented by Nature with a neat demonstration or 'physiological preparation' of the mechanism by which *functional* action proceeds within the family.

Let us look carefully at the construction of this placenta in the wall of the womb. Instant upon nidation, or nesting, of the newly fertilised ovum, the contiguous cells of both ovum and mother begin rapidly to divide, each contributing a membrane of its own building to the formation of the placenta. Thus mother and embryo together build the work-bench at which they are to work together as master and apprentice for the accomplishment of the pregnancy. In the depth of this placenta lie the lacunae or maternal blood-pools, rich storehouse from which 'the pregnancy' henceforth is to draw the material for its joint building programme. The placenta is gland-like in structure, and all that passes between its two very individual selective surfaces is conditioned to the needs of each of them in consideration of the needs of the other—a concrete, visible and tangible example of mutuality of organismal synthesis.

Here in pregnancy is no parasitism, but the most perfect mutual contributory relationship of which we know. No better 'demonstration' could be devised for the study of function; no more perfect exposition be found of the mechanism of mutual synthesis than this one of mother-and-embryo. The placenta, the tangible organ which links them, is a type-mechanism whereby mutuality of action is achieved. It is interesting that in the evolutionary scale differentiation of the zone of mutuality between parent and offspring into a discrete organ should be found in mammals—

the most highly developed species of which Man is the supreme example.

By the time the placenta is formed, the tissues of the maternal body have already passed over into a more fluid or 'reversible' phase, similar to that found in the plasticity of the tissues in the childhood state. This increasing fluidity enables the essence of every organ in the mother's body, now richly bathed in her circulating fluids, to seep out into her blood which flows through the placenta. In this fashion the experience—physical, bio-chemical and functional—of the mother already well versed in body building[1] is placed at the disposal of the developing foetus, whose contact with that experience is tempered by the intervening mechanism acting as a *zone of mutuality* linking the two into one functioning unity. It is through this zone of mutuality that the foetus comes now to be nurtured on a new and rich store of substances hitherto unencountered. But, as we have seen, it meets these in a form already *familiarised* and thus rendered acceptable.

The foetus, too, has its contribution to make to the pregnancy. Through the glandular activity of the placenta its own specific secretions flow back into the maternal blood pools, there to stir new developments in the maternal body. It is, for instance, as a result of this process that at quite an early date in the pregnancy the maternal breast begins to change and develop, and the mother to acquire a further stage of maturity which will later enable her to carry on lactation. So foetal *ex*cretions once again become maternal *in*cretions which stir the maternal endocrine processes.

Many stories are abroad of the effects of pregnancy and of labour on the father, but nothing is known of this matter nor has any experimental investigation been made on the human subject.[2] But in other species there is already a considerable body of facts concerning the physical changes in the male coincident upon mating and parenthood. At this point it might be helpful to digress and state at some length something of what is known about this subject.

[1] Not only has her own body been brought to adult state, but all its cells are constantly being replaced to make up for wear and tear.

[2] Experiments were under way at the Centre, before war brought our work to a close, to find bio-chemical evidence in the father of parenthood occurring in the family.

Some interesting observations have been made in the case of doves.[1] Incubation of the egg takes eighteen days, during which period both birds sit in turn upon the nest. It appears that about four days before the eggs are due to hatch, there devel ps in the crop of both female and male bird a 'crop' gland. This gland secretes 'crop milk', an enzymotic fluid which pre-digests, or, as we should say, 'familiarises', the food regurgitated for the newly-hatched chicks. Once the eggs are laid, the male bird thus not only seems to take an equal part in parenthood but, as a result, to undergo in his body similar *physical* changes to those induced in that of his mate. Important, too, in this connection is the fact that the chick deprived of crop milk in the early days after hatching cannot be reared on non-predigested grain. It, too, is dependent for its existence on the full functioning of parenthood. Here, then, in the doves there seems to be evidence of actual physical change induced not only in the mother and the embryo but in the whole *family-organism*—male, female and offspring—through parenthood.

The sequence of events in the doves is, however, even more interesting as an illustration of the family significance of parenthood. If, during the course of incubation, the eggs should chance to become addled, both birds still sit for the full 18 days (for that seems to be determined by copulation), but in this case no crop gland develops in the neck of either bird. The impetus to 'sit' is, therefore, independent of the quality of the egg once laid, but further development in the parents does seem to depend upon the viability of the egg, for death of the embryo interrupts the progressive mutual synthesis in the parents. It looks as though something—perhaps change in temperature, perhaps passage of some at present unknown substance of an endocrine order—is transmitted from inside the live egg to the parents, promoting their development.

One last point. It is now known that a male bird which has never been in contact with either a female or a nest of eggs can be made to become a sitting bird and to develop a crop-gland after injection of the buccal secretions, (i.e., saliva), of a sitting female. From this we can see that mutual synthesis of the organism is not only affected through the developing egg but

[1]See especially the work of O. Riddle and colleagues. Carnegie Institute Station for Experimental Evolution, Washington.

also by direct male and female contacts—the natural process of 'billing' and 'cooing'. We have evidence, then, of something which we can call a 'functional organisation', shared by the doye family—even if there is here no tangible organ like the placenta shared by the mammalian mother and foetus.

Subtle indeed are the influences that promote function in the organism! We can but presume that where humans are concerned, we are as yet only on the fringe of a most profound subject. It is not then without any basis for doing so, that we shall expect, given suitable circumstances, to find evidence of spontaneous development in the whole human family—father, mother and children alike—as a result of parenthood.

The closely co-ordinated process of nurture carried on under autonomic guidance does not end with foetal life. During pregnancy the mother is already preparing for the next stage of transition at birth. How does she ensure that the next food shall be familiarised for the child to be born into so strange a world? She provides a new food—breast milk. This milk is made from the *same* blood that fed the embryo, and which thus is allied in all the intricacies of its composition and in the uniqueness of its basic design to food which the learner already knows. Once again we see clearly the principle of *familiar nurture* in the carry-over of specific elements from an already familiarised environment—the womb—to one of a wider familiarity—the nest or family 'hearth' at birth. That unique individual, the child, under Nature's guidance is thus not only born of a specific family but also *consistently nurtured in the uniqueness of both substance and quality of the family of its origin.*

With the birth of the child into the family circle or nest, mutual synthesis does not cease. Even lactation is no one-sided transaction as between mother and infant. It is a process through which the mother casts off her transitory provision for the pregnancy and, conditioning her organs to a more matured state,[1] acquires a new and maternal configuration.[2]

It is the union of the pair that has set in motion this truly phenomenal train of events in the family organism. Is it likely,

[1] Lactation is known materially to assist in the process of involution of the uterus after childbirth.

[2] It would appear that it might take more than one pregnancy fully to confirm the maternal configuration. In the common parlance of the farmer, the dairy cow remains a 'heifer' until her second calving.

considering the compulsive behaviour seen in animals and the subtleties of close physical association in parenthood, that the male influence ceases with the birth of the child? Perhaps the old wives' tale that the man's anger, or his non-cooperation, can sour the milk is not so utterly unlikely! When we recall the fact that a functional unity can be maintained in the face of a dispartite organism—as in the ant heap—there can be no difficulty in envisaging the continuation of potent functional influences in the family, operating on all its members alike.

In this clear-cut procession of events in the early development of the individual which we see unfold under the autonomic guidance of the organism, the significance of parenthood becomes explicit. Three features appear as crucial:—(1) that *mutual synthesis* in the family leads all its individual members, parents and children alike, to move together towards further development; (2) that development of the offspring is brought about through *familiar nurture*, the new individual being led, coincidently with the parents' own development, from one familiarised environment to the next within an ever-widening family idiom; (3) that development proceeds through mutual synthesis carried on through a functional *zone of mutuality*.

We have watched Nature at work in realms beyond man's power of intervention. We have seen that in the course of the parents' own development after mating, through a continuous process of mutual synthesis the offspring is provided with the environmental circumstances and the nutriment that it needs. We have seen too that the particular significance of parenthood lies in the power to sensitise or 'familiarise' this nutriment for the offspring for each new stage in its journey.

It is with no uncertainty that Nature has indicated to us that not only is it parenthood which creates the new and unique individuality, but that the father and the mother are *specialists* for the specific nurture of *their* child. Parenthood is in fact the biological process evolved by Nature for the *rearing* of the young as well as for their initial creation.

CHAPTER III

BASIC TECHNIQUE

The power of science lies in the knowledge it gives enabling Man to bring his actions one by one into conformity with natural law. It is, for example, through a knowledge of, and strict conformity with the laws of aero-dynamics, that he has won the freedom of the air. Already biological studies have removed him from a position in which—considering himself born of a supernatural act of creation—Man singled himself out from the rest of creation, believing himself subject to some 'higher law' to which implicit obedience was also 'blind'. The work of Darwin cut the strings of this puppet-like suspension above the stage of the rest of creation. Now we recognise Man as but one of the species undergoing evolution in cosmos.

This release has brought with it the realisation that Man also is subject to natural law; it has freed the biologist to make search for the laws that govern his living with the same confidence that the physicist set out in search, with such success, for the laws that govern matter and motion. Observation and experiment are the method of procedure. Are we in a position to carry knowledge of Man's living a stage further?

From the two preceding chapters we have seen that the living organism is inseparable from its environment; we have seen also that to observe organismal function in the most discretionate form we must study Man:—*homo sapiens*. From observation of the autonomic or involuntary activity of the organism, certain processes carried on with consistency arrest attention. Making use of these as clues, can we design an experiment in which through the conscious or 'voluntary' life of the family, the operation of similar or identical principles underlying its functional action could be revealed? It is with this purpose before us that as human biologists, we set out.

For such a study there are certain pre-requisites:—

1. The 'unit' of living material for study must be 'the family' in its biological setting.
2. Dealing as we propose to do with volitional action, the experimental circumstances created must be such that

the unit under observation is free to act voluntarily rather than in conformity with any pre-determined conduct, in pursuit of any ideal or in response to any external discipline.

3. The environment must contain a maximum diversity, so that there may be adequate chances for the unit under observation to exercise its volition, and for its biological potentialities to become explicit in the ordinary circumstances of living.
4. There must be at least a minimum aggregate of units to provide the requisite social contacts permitting diversity of action by the family, as well as providing statistical data for the scientist.
5. The units must be in a position to assimilate, as part of their natural environment, the technical organisation of the scientists undertaking the observations.

The material for study is to consist of families. The 'family' we refer to arises with the mating of two specifically diverse individuals, developing as a 'unity' into a functional organisation. Thus it is not the family of the geneticist with which we are concerned—that hereditary entity represented by the genealogical tree ; nor is it the social entity derived from the same source and often consolidated by little else than the tenacity of convention or inheritable possessions.

The laws governing the biological needs of the family being as yet unknown, the family as a functional organisation receives but scant consideration. As a social aggregate it is, perhaps, legally recognised as an entity only by the tax collector, the public assistance officer and the relieving officer. In modern social organisation we are accustomed to take account only of the *individual*, about whom all the activities of daily life regarded as important are designed to revolve—e.g. industry, economics, politics, sickness and welfare administration ; and also education. Even a matter so closely connected with family life as housing is in most cases determined, not by the necessities and the potentialities of the family as an organism, but solely by the applicant's ability to pay the rent. The mere fact, therefore, of basing any organisation on the family-organism as a unit, implies a new and unique orientation in modern society.

It is to experiment with the family in this sense that we are concerned in this book. What families are we then to select for the purpose ?

In order that any studies made shall have general validity, the families collected must be as representative as possible of the general populace. But the nature of the proposed study imposes many conditions upon us in making our choice.

As we have already seen, organism and environment are inseparable. So it is families in their natural habitat or everyday setting in which they grow up and live their ordinary lives, that have to be sought for study. Nor is it families in isolation, one taken from here and another picked from there, that can form the selected group, for they must be so aggregated that they may act in mutual synthesis with each other.

The next consideration to be taken into account is that the public from wherever selected will inevitably consist of a mixture, some of whom are sick and hence, whether wittingly or unwittingly, obeying the laws of pathology. We shall not look for health among the sick ; it is the presumably healthy moiety that alone can serve for this enquiry. But since we have no means of knowing in advance which these are, and still less of effecting a natural separation of them from the pathological, at the outset it will be essential to choose our sample with some care, selecting to the best of our extremely limited knowledge the most healthy, as well as representative cross-section of the populace. This entails the avoidance of any group drawn from a social-problem class, in which disability, whatever its origin, hampers the expression of potentiality ; the unemployed, the unemployable whether rich or poor whose tendency to flock together is common knowledge. Choice must lie rather with a sturdy sample of society ; families not taken from the all too pleasant backwaters nor from the stagnant verges of the stream of life ; families neither too heavily laden nor too lightly freighted, but those who are in fact good swimmers in midstream.

It is also important that the sample of society chosen should not be composed of an aggregate of families selected for any purpose—e.g. all very young families which the biologist might well be tempted to select—for, by ourselves collecting and bringing them into association, we should by external suasion be causing their segregation, imposing our own bias upon them and disturbing such social relationships as they had formed; we should in fact be taking them out of their natural habitat. Nor would it be satisfactory to recruit families on the basis of some idea or ideal: e.g. the Church, enthusiasm for music, for politics or sport or any other single interest, because, as we shall see later, we require diversity of every sort in the sample of the populace chosen.

Then, again, the nature of the approach that we propose to make to the families will also effect the selection to be made. The usual method for the collection of data and information by any student or investigator of sociological problems is by invasion of the intimate environment of the individuals to be studied. The biologist is not in a position to proceed in this fashion. He, requiring equipment and instruments under his direct control and manipulation, must extend an open invitation to families to come to him, bringing with them their own environment.

We need to be in a position to observe the family in action. The three main spheres of activity of the adult populace are: industry in the case of the man and the unmarried adults of both sexes; the house in the case of the married woman; and leisure, common to all. The first of these is not orientated in relation to the family; the second is a sphere which, being of a *specific* nature, from the biological point of view cannot be invaded with impunity. Neither industry nor the domestic hearth, therefore, can provide the required material. So, in the first instance, it will be to the leisure of the populace that the biologist must turn.

The laboratory of the biologist will have to be within the field of the leisure of the family and must be so constituted as to make continuously available all those things people naturally do in their leisure hours. Besides affording a focus for the spread of knowledge, it must contain all the essential means through which the family may make the social contacts from which action naturally proceeds—such as sport, games, dancing, reading, music, drama, etc.: and these must be available in as great a diversity as practicable.

Apart from such diversity of material factors, the environment of the families under study must also contain a diversity of biological factors. To achieve this, the first necessity is a sufficiently large aggregate of families—probably not less than

2,000 families—in order to permit of an adequate cultural admixture. But an adequate cultural admixture cannot be derived from those closely segregated into one class, one wage level or from those working in one close industrial preserve. No new housing estate consisting of families of one social stratum, no area in which the working members of the families operate in one single industry, and no group representing one level of culture, would yield suitable material.

This requirement of diversity, however, is not fulfilled merely by the cultural variety of the families that are to be assembled. The necessary biological diversity will also be derived from the variety of action appropriate to every stage of development, so that *varying stages of individual maturity* also form a necessary component of the diversity of the environment. The fact of assembling families, of itself provides this type of diversity. An aggregate of families gives automatically both a *vertical* and a *horizontal* grouping of every stage of development.

Furthermore, all factors contributing to diversity in the environment, whether objects or actions, must avail in continuity, so affording the possibility for all members of each family to make frequent and repeated contacts with each new experience as it becomes pertinent to their own development. By this means the organism will be enabled to exercise its growing power to digest new material at every step. So continuity in the association of families chosen is another necessity for their development. Only a more or less closed geographical zone can provide families between whom contact would be likely to be maintained in continuity. Hence we are compelled to make the site of our operations a local one.

The circumstances created must enable contact to occur spontaneously in some social meeting place where families meet naturally and in freedom in their leisure hours; some place where a sufficient number of families can foregather in a social milieu and interact one upon another while making use of any chances for action present in the environment. So the situation we have to envisage is one in which families under observation may find at hand material of many sorts—things, people and their actions— as and when they evince a desire and/or ability to use them; i.e. to enter into a relation of mutual synthesis with them.

Moreover, the environment must be and continue to be sufficiently fluid to enable those factors to be made use of, not necessarily in the conventional and accepted way, but according to the growing needs and capacity of each family as it develops. Merely to provide a glut prior to the appearance of any evidence that it is utilisable by the family, would be to provide an environment in which the inept might well wilt—like the axolotl taken out of its pond before metamorphosis is imminent. Nutriment for the family in the absence of the power to utilise it is as good as useless. It is with the *power of family utilisation* that our experiment is concerned.

So, it is not merely with the physiological competence of the machine that we shall be concerned in human biology. It is with the mutual synthesis of the family and its environment including experience of every type—physical, mental and social. In all these spheres environmental diversity must be attained, for unless the environment does contain that which will afford appropriate nutriment for the next stage in development of functional action—for example a bicycle to a child ready to explore his faculty for balancing ; association with adult society to an adolescent reaching maturity—development must proceed unbalanced, or be arrested. Not that a bicycle, even were it always available, is at all times an opportunity to any child ; it is converted into an opportunity at the moment when capacity for the achievement of balance through bicycling has reached a critical phase in the development of that particular child. So, too, with adult society to developing adolescents —each acquires importance for them in due season. It is the relation of the circumstance to the individual's state of development—its topicality to him—that converts *chance* into *opportunity*.

We have seen that in educating the foetal and infant learner in the power to utilise new experience, the parents under autonomic guidance proceed by the method of *familiar nurture* to pass him by gentle stages from a closely specific and familiar situation to one of a wider specificity within the sphere of parental function. To see whether, conditions being favourable, this same process will be carried on naturally throughout parenthood, circumstances must be secured in which ever-widening experience of all varieties may pass through the parental mill for the feeding and rearing of the child. So, as he approaches adult stature we may hope to see him reach his full and specific individual maturity, before himself forming a new family-nucleus. It is in fact a slow method we must envisage in preparing families for biological study ; there is no short cut to the evolution of human function.

We now come to a very crucial question :—how can such an unfamiliar and objective factor as a scientist and observer be introduced into any social milieu without instantly shattering its spontaneity ? The answer seems to lie in the possibility that the scientist himself and his technicians should become one of the accepted groups forming the cultural diversity in the environment. Fortunately, as biologists concerned with function we shall not necessarily be called upon to make investigations into the genealogy of the family, to look into the pretexts for its social prestige, to examine its credit or make other studies distasteful to its members. This clears away at the outset many difficulties of approach which beset other workers on human material, laying them open to criticism as intruders or even as busybodies—e.g. the eugenist, the social worker, the economist, the psychologist, the clinician, the preventive therapist.

The biologist in this experiment, being a person requiring as his technical instruments the appurtenances for every sort of activity for the leisure hours of the family, has at once a mutual basis for association with the public that is his chosen material. We can begin to visualise the situation to be created ; it is one in which both biologist and families share the appurtenances of leisure. And here we must hasten to say that such a relationship will imply a special discipline in the observer. He must learn to rid himself of all preferences, of all pre-conceptions as to how and when and why things should be done. The attitude of complete impartiality so admirably acquired by the doctor dealing with physical and psychological disorder in the consulting room, must be extended by the practising biologist to social disorder and unsocial behaviour in the family. All these pathological conditions, which he will inevitably meet in his daily work, must be met without prejudice—and not merely in the confines of the consulting room, but in the circumstances of everyday life. It is clear that for this no commonly accepted social—or medical—training and no short probation will suffice.

Through the use of the same material, each for his own individual purposes, the scientist and the families that are to constitute the object of his study acquire a mutual point of contact in experience. Nevertheless there is an important difference between them, for while the biologist is educated to evaluate the actions of the individual, the individual is not yet educated to evaluate the actions of the biologist. With this point we shall deal later.

The scientist's purpose in approaching families is to be in a position to make a biological assessment of their functional action. This will involve :—

(1) Observations of the actions of the individual and of the family as a whole in response to the flux in their environment.

(2) A study of the changes that ensue thereon in the environment itself.

(3) Physiological studies of each individual and of the family who are thus acting on the environment.

Can the knowledge that the biologist gains from this source become of use to the families ? Knowledge of every kind is one of the most important factors contributing to the development of the individual and hence to diversity in the environment. How often, especially among skilled wage earners, one comes across the highly sensitive, alert and intelligent man who, because deprived throughout the course of his life of general knowledge pertinent to his own special interests, remains inarticulate and unable to implement his gifts in any sphere beyond that of his circumscribed intimate and personal one. Diffusibility of knowledge throughout the environment in which the families are to move is essential if the full expression of their potentiality is to become explicit in action. Facts pertaining to experience of every sort that the family is in course of digesting give the context and the full flavour of consciousness to their experience. So knowledge must be at hand and readily communicable from family to family. This condition can only occur in an integrated society where it can impinge upon the family organism in all the various phases of development and in all the vicissitudes of fate and fortune.

In an association of families such as we are looking for, there will be three possible sources of knowledge continuously available ; the subjective knowledge that the individual may derive from his own actions ; the objective knowledge derivable from his interaction with other individuals and families ; and the special knowledge derivable from the biologist's three spheres of observation already mentioned. Not the items of this special knowledge, but its significance to themselves is something each family would like to have and be able to use. Can the biologist when acquiring his facts return them in any form utilisable by the families for their own use ? This will demand the development of a new technique and much patience, but it should be possible and to the mutual advantage of observer and family alike.

What a strange laboratory it will be that fulfils all the needs we have postulated ; that will use human families as its material for study ; that will enable the spontaneous evolution of a freer and more diverse environment for those families ; that will make possible the development of latent potentialities of the family as a whole ; that by the nature of its constitution, will permit of the familiar nurture of the child throughout its development up till adolescence ; and that will afford the spontaneous association of biologist and family, as well as of family and family in their daily coming and going. It cannot be any sheltered or secluded spot in which parts or fragments of the family organism can be isolated, dissected and analysed. It must be an open field upon which every influence may play free as the changing winds, upon people of all ages and all sorts, the observers themselves being of the company and functioning in unity with the whole.

What a task for any architect to plan a building intelligently and usefully for the general purposes of family leisure, the building to carry a mixed cargo of all ages, both sexes and of all interests —and not in any haphazard way, but designed to meet the needs of families as they grow ; and besides this to afford an observatory in which all action unfolds before the eye of the scientist moving about the building.

The scene of action, intimate and popular, will be very different from any laboratory hitherto in use, so that, to the casual visitor, its real purpose may escape recognition. This is in fact what did happen to the first laboratory of this sort, the Pioneer Health

Centre in Peckham, which has often been mistaken for a Welfare or Medical Institution, or confused with the polyclinics for the treatment of disease mis-called 'Health Centres'.[1] More often it has been taken for a Social Club of a rather elaborate design, or even for a pure recreational or amusement centre for the frustrated and the bored. Only as it has developed and as its intentions have become explicit through action, has it finally come to be widely known as "the Peckham Experiment"—a new venture in the science of human biology.

[1] The Pioneer Health Centre was established in 1926. All subsequent 'Health Centres' known to us have been in the nature of Polyclinics, in which are assembled not only Maternity and Child Welfare Services but also many provisions made by the Local Authority for the care of the sick. Thus the presumably healthy mother and her new-born child are deliberately aggregated in the same building, for instance, as children recovering from the extraction of tonsils, with those attending clinics for the tuberculous and not uncommonly those suffering from venereal disease. The fact of the mothers and infants entering by a separate door does not expunge the psychological implications of assembling the (presumed) healthy among the sick.

See also the Draft Interim Report of the Medical Planning Commission (B.M.J., June 20th, 1942, p. 749) in which it is proposed that the pregnant mother and her child, both presumed to be healthy, are to attend a 'Health' Centre which is still in effect a 'surgery' largely concerned with the treatment of the ambulatory sick.

ACKNOWLEDGMENTS

The authors wish to express their grateful thanks for permission to reproduce the following photographs :—

Architectural Review :	Fig. 1
Dorien Leigh :	Figs. 2, 3, 5, 6, 9, 13, 14, 15, 24, 28, 39, 42, 45, 46
Harry Grose :	Figs. 33, 41, 43
F. R. Reed :	Fig. 44
Star :	Figs. 22, 27
Topical Press :	Figs. 11, 16

THE CENTRE LIFE

Lest the reader should imagine that this book is an abstruse dry as dust record of scientific experiment, we open the chapters describing what has been accomplished with a pictorial presentation of family life in the Centre.

The first thing that struck most visitors was what they usually described as the 'atmosphere of the place', commenting on the forthrightness of the members—adults and children alike; and, with surprise, on the absence of any self-consciousness in the people gathered there.

The biologists' laboratory—a family club *Chapter IV*

The weekly subscription covers the whole family

The café round the swimming-bath

Chapters IV & VII

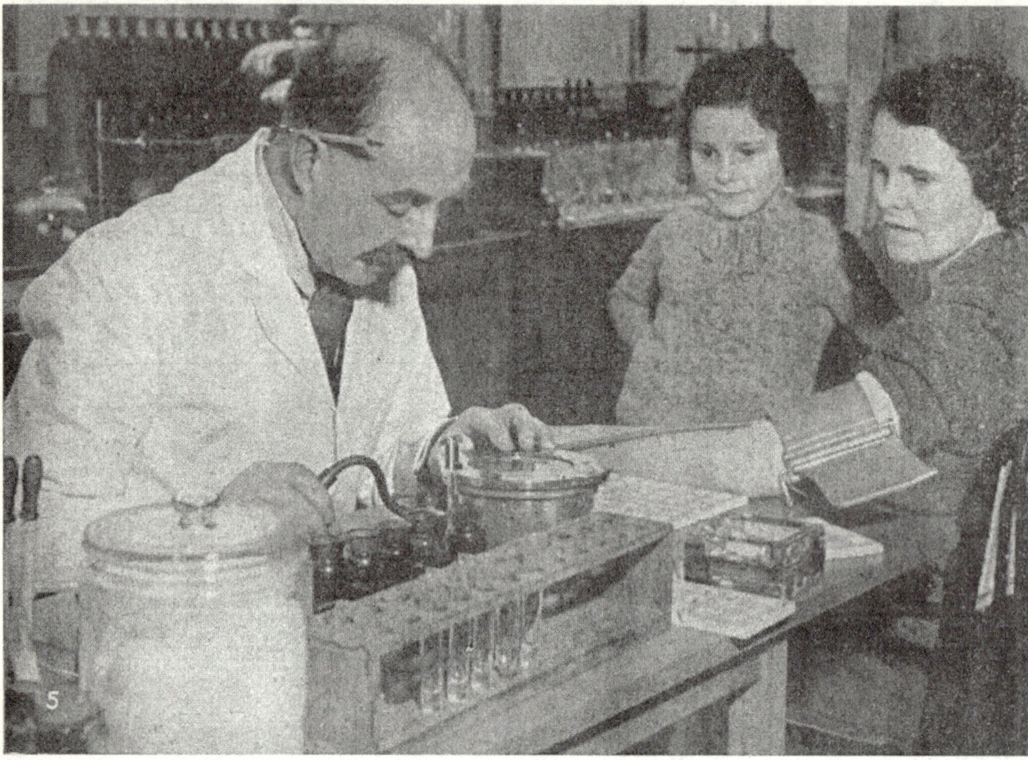

Biological accountancy *Chapter V*

Biological budgetting ; the 'family consultation' *Chapter V*

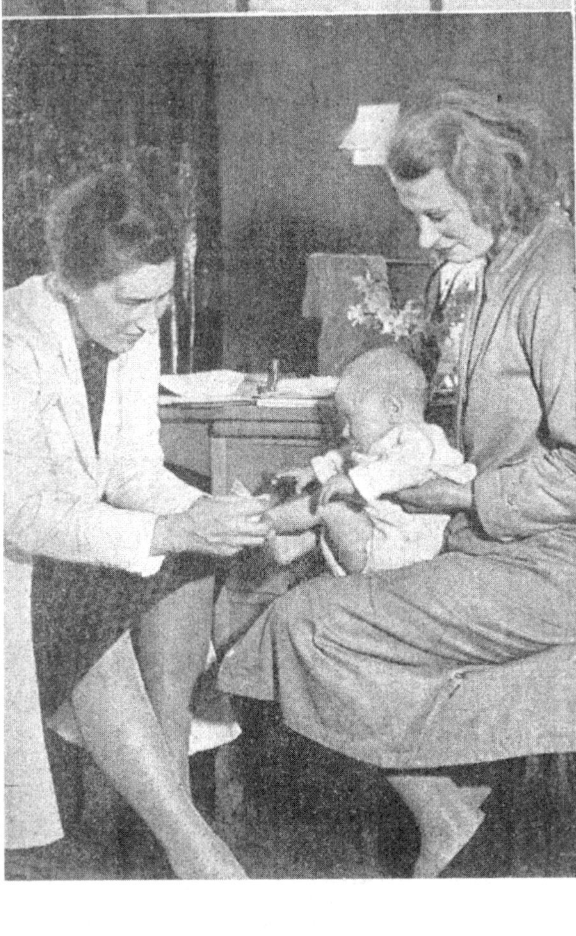

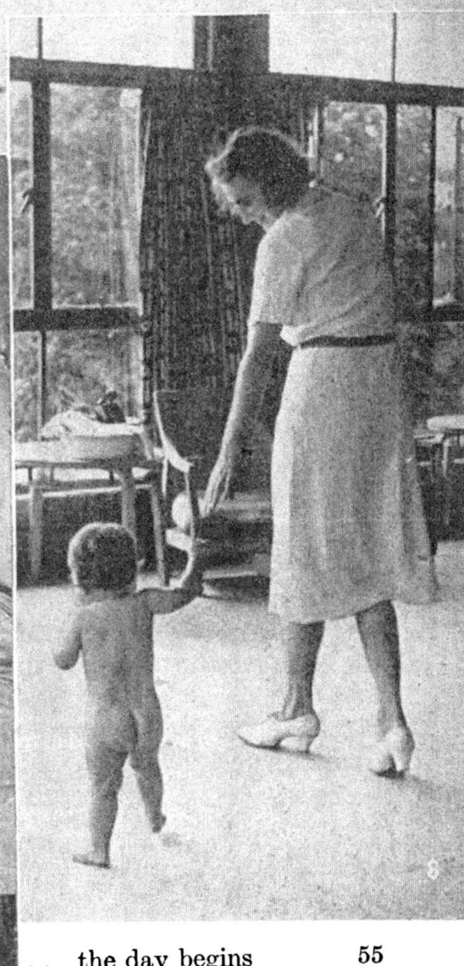

.. the day begins

Chapter IX

Weaning the child—*and* its mother *Chapter IX, section 2*

Self-service not an expedient but a principle *Chapter IV*

Taking his place in the social circle *Chapters IX, XIV, XVI*

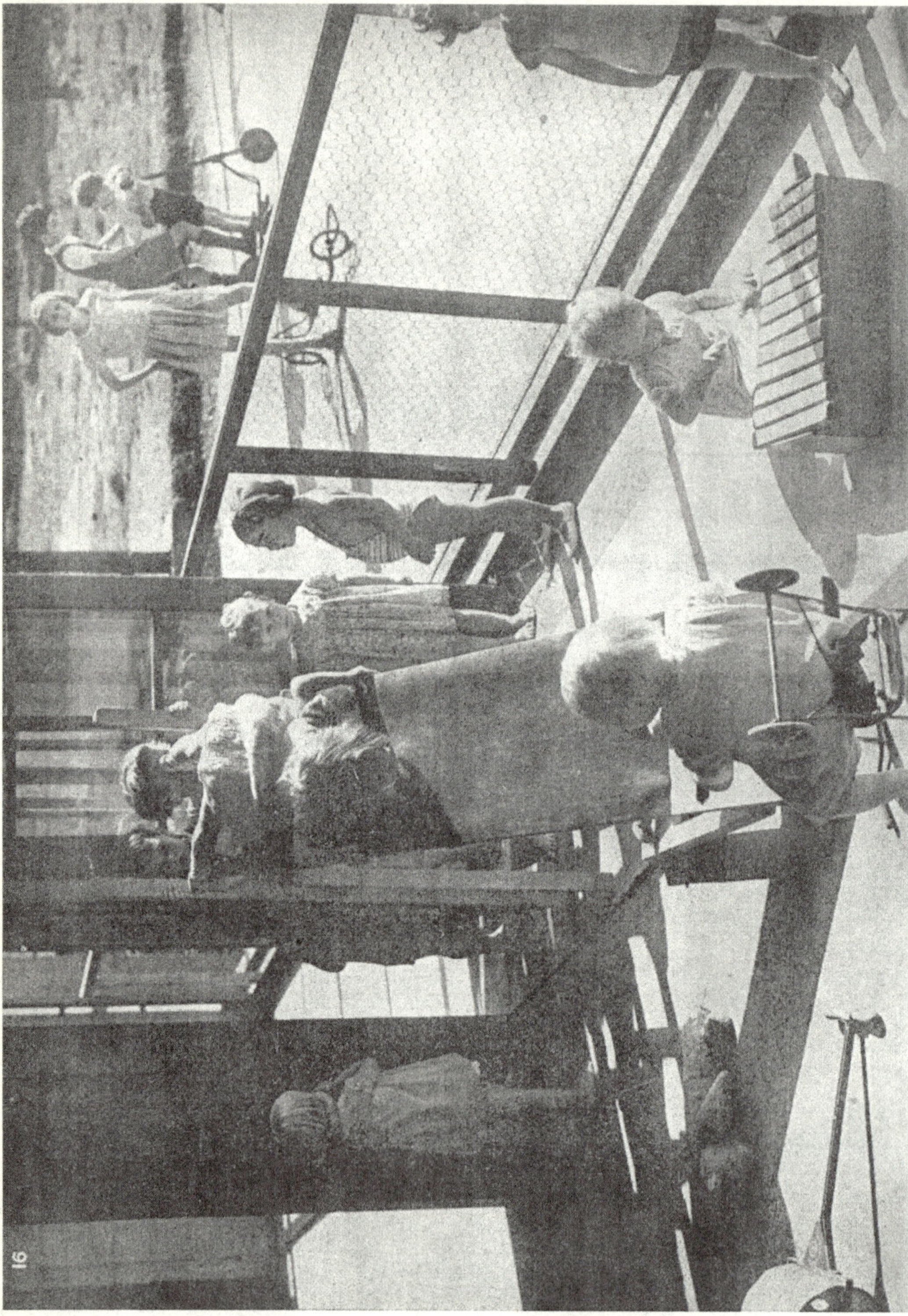

Their own place in the sun *Chapter IX. section 3* 58

discovering their faculties
59 *Chapter IX, section 3*

Released from school they move freely through the society of the Centre; 60
find the 'curator' and choose what they will do . . . *Chapter X*

. . . free and unafraid *Chapter X*

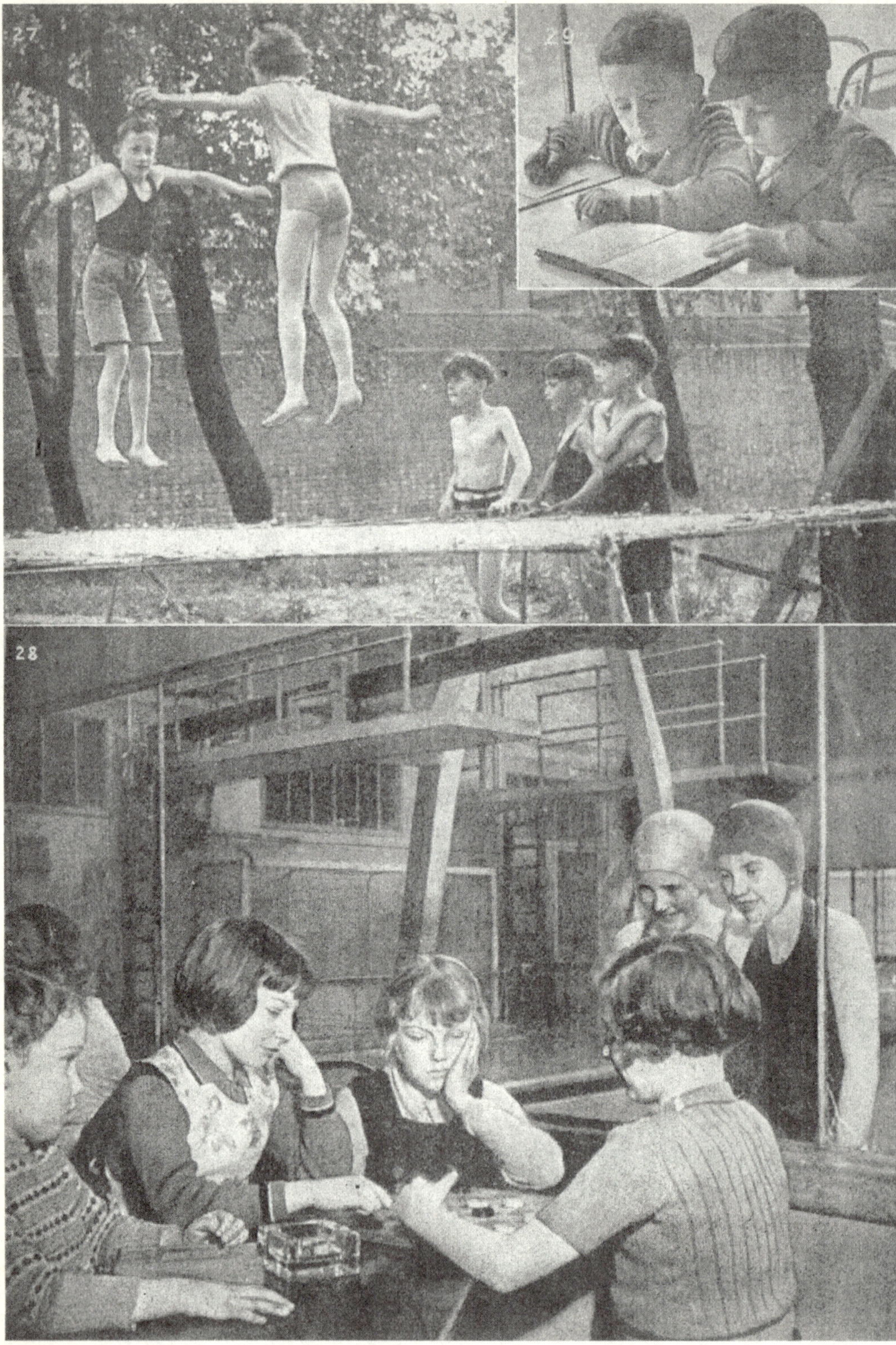

intent on skill . . . *Chapter X*

30

31

32

33

Skill and social integration grow together *Chapter IX* **63**

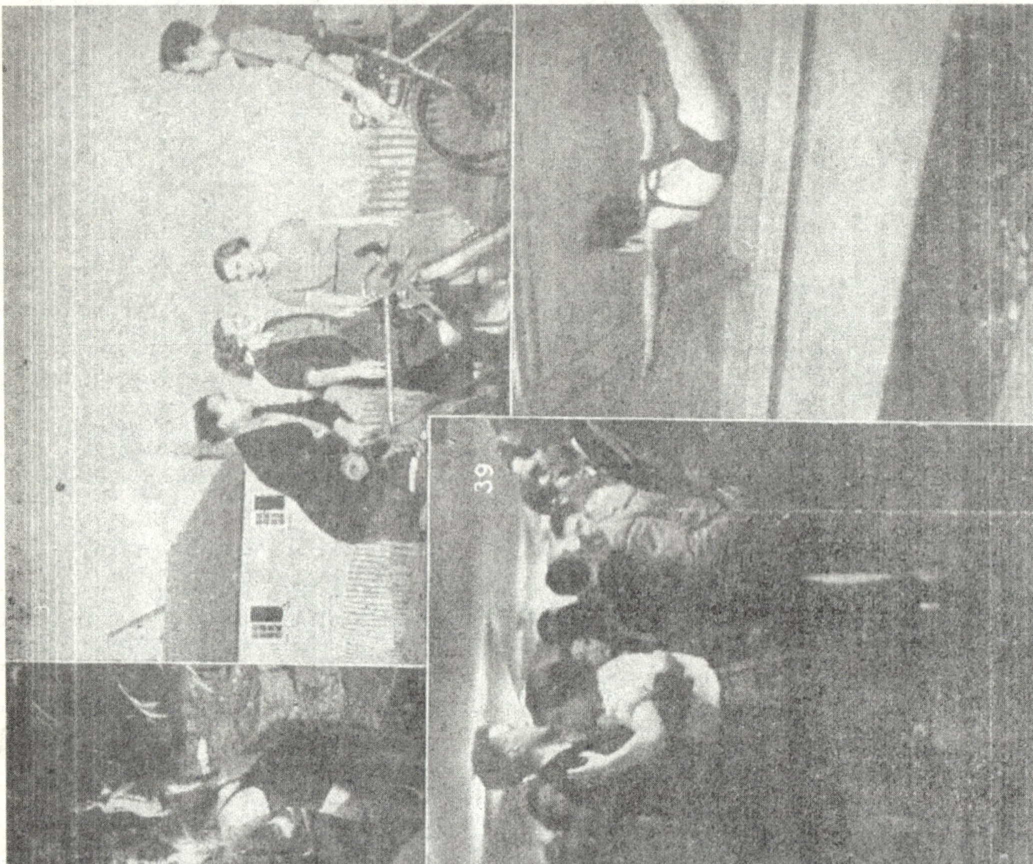

Growing up—in a world full of opportunities *Chapters IX & XII*

41

42

43

44

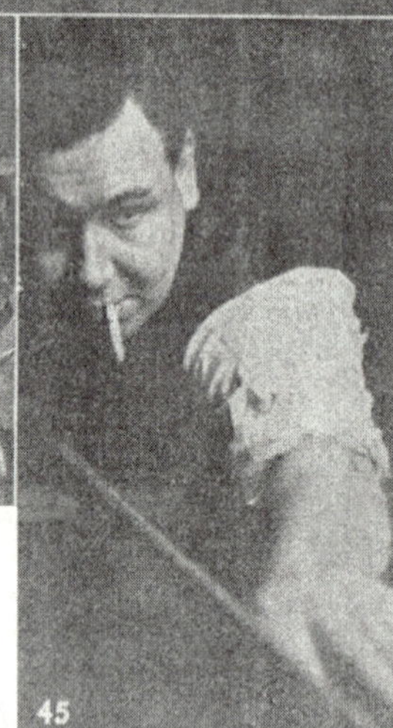

45

Saturday night—among friends *Chapters VII & XVI* 66

46

CHAPTER IV

THE HEALTH CENTRE

The Building.

SET back 100 ft. from the pavement of a quiet street, only a stone's throw from a main thoroughfare in South London, stands the building officially called the Pioneer Health Centre, but by those who use it, more familiarly known as "the Centre". It consists of three large concrete platforms (160 ft. by 120 ft.), rising one above the other cantilevered widely over supporting pillars arranged in parallel series, surrounding a rectangular central space occupied by a swimming bath (35 ft. by 75 ft.). This form of construction on pillars allows the outer walls and those of the centrally placed swimming bath to be of glass, as indeed are nearly all the very few partition walls within the building. The front with its series of bow windows, the sections of which fold back in summer, presents to the eye a structure of open balconies one above the other, designed to be colourful with climbing creepers and to catch the afternoon sunlight.[1]

Along one side of the top floor is the only space that is shut off from general circulation. It is a Consultation Block, consisting of private consulting rooms, reception rooms, changing rooms and bio-chemical laboratory. Except for its very frugal and light construction, this block has the more or less conventional appearance of a medical department of any modern clinic. The remaining part of the top floor consists of large light open spaces for quiet occupation, library and work-room, games, etc.

The whole of the floor below, i.e. the first floor of the building, is taken up by a cafeteria and by a large hall for social purposes, from both of which the central swimming bath is visible through a continuous encircling band of glass window. From the long hall, looking down through two large windows on to the ground floor, are seen at one end a gymnasium and at the other a theatre. The rest of the ground floor consists of infants' nurseries opening on to the grounds ; of an infants' and learners' swimming bath, which again can be seen through a window from the passage leading to the nurseries ; of cloakrooms, changing rooms and spray chambers, etc., for the bath, gymnasium and

[1]See photograph 1, p. 52.

theatre. The land in front of the building, apart from an area of concrete used for roller skating, cycling, etc., remains largely in the rough awaiting development of the experiment to give direction to its lay-out.[1]

Here then is a modern building designed as a laboratory for the study of human biology. The general visibility and continuity of flow throughout the building is a necessity for the scientist. In the biological laboratories of botany and zoology the microscope has been the main and requisite equipment. The human biologist also requires special 'sight' for his field of observation—the family. His new 'lens' is the transparency of all boundaries within his field of experiment. Sixteen steps down from the consulting room and he is engulfed in the action that is going forward, and which, by reason of the very design of the building, is visible and tangible to his observational faculties at all times.

Apart from the architectural lay-out already described, what equipment was there in the building when it opened ? A few books in the library, a billiard-table, games for the children, a gramophone, one or two pianos—nothing more. All the activities which later we shall see taking place in the Centre have been added one by one as the desire for them has arisen out of the association of the families gathering there.[2] It may be difficult in the first instance for the reader to conceive of such a building ; difficult to rid the mind of a picture of a glass-house with large open spaces in which little comfort and sociability would be likely to forthcome. To those who use it, it does not appear like this. The main social floor looking on to a blue-green swimming bath, the ripples of water casting their ceaseless reflexions on walls and ceiling, has at once an attraction that is unique. On the sunny side the building has gently curved window-bays and everywhere the large open floor spaces are broken up by pillars which, besides giving variation of light and shade, give a number of foci in the aisles about which intimate groups can form naturally without interfering with the continuity of circulation throughout the building. So that, although there are no culs-de-sac and no parts of the building shut away, there is no sense of living-in-a-passage nor of uninviting bareness.

[1] See Appendix I, for sketch-plans of and notes on the building.
[2] For a list of the activities that have grown up in the Centre, see Appendix II.

To appreciate the salient features of its planning and purpose, it must be seen in use, for it is essentially a building *designed to be furnished with people and with their actions*. There for young mothers with time to spare in the afternoons, for infants ready for adventure, for school children to come to when school is over, for the adolescent as well as for both mother and father and all grown-ups when work is done, it is, too, a place where the family can foregather.

The whole building is in fact characterised by a design which invites social contact, allowing equally for the chance meeting, for formal and festive occasions as well as for quiet familiar grouping. It is a field for acquaintanceship and for the development of friendships, and for the entertainment by the family of visiting friends and relations. In these times of disintegrated social and family life in our villages, towns, and still worse in cities, there is no longer any place like this. Nevertheless, man has a long history of such spaces that have met the needs of his social life and the tentative adventure of his children as they grew up :—the church, the forum, the market-place, the village green, the courtyard ; comfortable protected spaces where every form of fruitful social activity could lodge itself.

The Centre is just such a place, not modelled on the past, not traditional, but planned to meet certain biological necessities only now beginning to be understood. It is an open forum, with its recesses roofed and warmed in winter, yet open to the sunshine when it comes ; a place where at all times any member of a family can drop in alone for an hour, or foregather for a party with equal ease and suitability—not a lounge for casual contacts, but an arena for the unfolding of the consecutive and integrated leisure activity of families.

The plan is based upon the principles of *biological cultivation*. Action in the building is not to result from any professional leadership, but to arise spontaneously out of the circumstances of the environment freely impinging upon the families as they use it. So all activities, sequestered though they may be because of their intrinsic nature—e.g. the swimming bath, the gymnasium —are at the same time visible to all who use the building for any purpose. The swimming bath illustrates admirably this dual necessity in construction. It is in an enclosed chamber, the temperature and humidity of which can be controlled, but through its encircling band of glass it is visible to the occupants of the rest of the building, and it is sight of action going forward in the water that will constitute the familiarising factor stirring the spectator to new achievement and drawing non swimmers to the attempt. So with the nurseries, the dance floor, cafeteria, theatre, library, workrooms, etc., etc.; all are planned to come within the vision of families and observers alike as they use the building.

The District.

What families are they who use this building? After careful preliminary investigation for a suitable district the building was erected in Peckham, in the Borough of Camberwell in South East London, and in May, 1935, was opened to the use of any family living within an area of not more than a mile in extent from the Centre: that is to say within *walking distance*. This locality was selected because of the varied types of family settled there: mainly artisans, occupied in every sort of skilled work, but also independent tradesmen, employers of labour, various grades of civil servants and municipal officers, clerical workers, some professional men as well as a few unskilled labourers. The income level of these families varies between £2 5s. 0d. per week and £1,000 a year or more, the large majority, however, earning between £3 10s. 0d. and £5 0s. 0d. per week. Thus varied in occupation and in income level, this populace roughly represents a cross-section of the total populace of the nation with as widely differing a cultural admixture as it is possible to find in any circumscribed metropolitan area. It is no sheltered group of workers but a sturdy people who have succeeded in keeping themselves in as fit a state as is possible in the face of present social conditions.[1]

The houses in the district, each with a garden behind, are mainly of two or three stories, built originally to house one family, but most of them now occupied by as many separate families as there are floors to the house. Little alteration has been made in these houses to adapt them to their latter day use, so that the accommodation is far from modern, and the sanitary arrangements inappropriate for the families they now

[1] For a further analysis of the populace chosen, see "Biologists in Search of Material," p. 55. See also Appendix III on nature of employment of Centre members.

serve. Living within our small chosen area there are approximately 5,000 families.

The building as we have seen had to be designed to accommodate the leisure activities of 2,000 families. Taking the average metropolitan family to consist of 3.75 individuals, 2,000 families represent an aggregate of 7,500 individuals. Time has not yet allowed us to reach that membership, but we already know from experience that fewer than 1,000 families in continuous membership do not provide the diversity of types and of actions adequate for the spontaneous emergence of a sufficiently varied culture. After four years' experience it still seems certain that twice that number is necessary, as originally postulated. But we have learned that the Centre as planned and built in Peckham is adequate in dimension and admirably suited in design for the use of the 2,000 families for which it was built. From the beginning we foresaw that any body of families linked in action as are the Centre members, would inevitably extend their associations beyond the confines of their club building. Thus the organisation would be bound, as a result of the growing activity of its members, to acquire extensions designed to meet the peak points of their leisure: half-holiday activities, week-end facilities, holiday camp, etc. This we shall see later has already begun to occur. Such extensions automatically relieve the pressure on the building when congestion is at its greatest. For all these, however, the Centre itself being the focus of initiation remains the integrating core.

Amidst the humdrum life of this residential district a new building has sprung up. The relatively few people who pass that way wonder what it is .... "They say it has a full-sized swimming bath, a gymnasium, a theatre, a cafeteria—where you can get beer"; dancing goes on there and moving figures can be seen on the floor of the main hall at night when the whole building is lit up attracting the attention of the passers by—if they have not already been led to pause by the strains of the band or singing filtering through the night air. What is the reaction of these passers by? Too often it is the normal reaction to novelty; they pass by without enquiry. There are relatively few questing and adventurous individuals in modern society, and this is a general characteristic, not one peculiar to any local populace. So the growth of membership of the first "Centre" is likely to be slow. Circulars announcing its opening and advantages will for some long time be little more than convenient firelighters. Perhaps a passing couple will see the notice board —"Queer sort of place that; you have to have a medical overhaul; wonder what goes on there?"—for they know few people in the neighbourhood and have never seen or heard of the place before. Or—"Let's enquire at the gate"—and at once an invitation follows to come in and look round and to ask for the Secretary on reaching the first floor. Once inside, the atmosphere strikes them as friendly; the process of familiarisation has begun. "Is it a Club? How do we become members? . . . ."

A Family Club.

The Centre is a Club for families,[1] admission to which can be gained by a family subscription of 1s. a week. The conditions and privileges of membership are two :—

(1) Periodic health overhaul for every individual of a member-family.

(2) Use of the Club and all its equipment, free to all children of school-age or under of a member-family, and by the adults on payment of a small additional sum for each activity.

These features are of a novel character. On first contact, periodic *health* overhaul is likely to be confused by the man with *medical* overhaul, of which he may have had experience in industry where he is by no means convinced that it operates always to his benefit. In the second place, here is an invitation to engage in gregarious action which cuts across the ordinary conventions of social life where most actions are carried on in segregation. How often this finds expression in the term—"we keep ourselves to ourselves"—an assertion made by members of specialised groups,—e.g. of a tennis or a bridge club, as well as by adults in general. But with the children no such barrier exists. They prevail upon their parents to accept the invitation that the post has brought them to visit the Centre. They have heard of it from their school friends, often watched what is going on from the road, and sometimes even managed to get in with a friend who is a member. When they come they miss

[1] Our definition of 'family' is the mated pair, with or as yet without children. A widow or a widower are also (biologically) accounted a 'family'.

nothing and never cease to pull and to persuade till their parents decide to join.

So by the end of a year and a half, we had drawn 10% of the 5,000 families in the district chosen.[1] In "Biologists in Search of Material" the personal and individual factors operating for and against membership have been fully discussed.[2] There are however two general factors of which we had become acutely aware towards the end of the fourth year. Even as late as this, families were continuing to join who, although they had lived in the district, had never heard of the Centre till the week before; people whose daily tracking was so consistently conditioned by circumstance and so closely confined by habit that, although they daily used the station 100 yards from the building, had never seen it and were unaware of its existence. The second fact of importance to membership is the constant shift of the urban population; here for a year, then gone to the next temporary halt, never long enough anywhere to grow any roots in society. From a survey made by us in 1938, it appears that of those whose membership of the Centre lapsed, 43% had left because they had moved from the district.

Certain other essential points in the constitution of the Centre and the technique used must be discussed. The subject of maintenance is one of the first questions asked by any visitor.

The organisation has been designed, with the achievement of full membership, to become a self-supporting one. The main sources of calculable revenue are two: first the weekly family subscription at 1s. a week per family, with full membership of 2,000 families yielding £5,000 a year; second, a further esti-

[1] Early in 1938 the number of families having joined and remained members being still inadequate for the conduct of the experiment, and being ourselves pressed for money, we were obliged to enlarge the district to include all families living within a mile in any direction. This subsequent area includes some 25,000 families. The results of this procedure cannot be given, since the war cut short the work before they could be adjudged. It was however apparent within a year that many factors other than walking distance operate to make membership possible or impossible; for instance, the presence of busy traffic routes across the path to the Centre may make it dangerous for children to come unaccompanied; the location of the shopping area in the opposite direction may make it difficult for a mother to visit both it and the Centre in the one day. The position of the school or of the station used by the workers, the tram or bus route—trivial though they seem—all operate to determine the frequency with which the building can be used by a family, and hence the duration of their membership.

[2] p. 34 et seq.

mated sum of 1s. a week net per family, derivable from the sum of small amounts paid by all adult members for whatever they do in the building—e.g. 3d. for a swim, 6d. for a whist drive, etc. —which should yield a similar sum ; these two main sources of revenue together giving an estimated revenue of £10,000 a year.

The unfamiliar nature of the enterprise to the populace, the experimental nature of the venture in which the requirements of research must take first place, and the relatively short time we have been established, all make it impossible yet to demonstrate the validity of this financial estimate. So far we can only say that the sum chosen for the weekly subscription is well within the capacity and acceptance of the wage-earners we cater for—for many, considering the sum fixed to be an inadequate one for the advantages accruing to them, asked to be allowed to pay more. The amount of extra money forthcoming as an average sum per week from each family we have found by experience to be within reasonable reach of the sum estimated. It depends on the rate of growth of the Club, for the more members there are, the greater number of activities are likely to be in progress ; and the more the Centre is used, the larger is the proportion of spare money spent there.

What of the outgoings ? For such an organisation to be self-maintaining a great deal of thought and ingenuity must be put into its initial planning with a view to reduction of the cost of maintenance. Some of the scientific requirements of the experiment have also a bearing on the financial aspects of running the Club. Of these, two illustrations might be given, the nature of the service, and the method of collecting and recording entrance subscriptions and other takings.

Self-Service Technique.

The 'self-service' aimed at throughout the buildings is a primary need of the biologist. A healthy individual does not like to be waited on ; he prefers the freedom of independent action which accompanies circumstances so arranged that he can do for himself what he wants to do as and when he wants to do it. The popularity in tube stations of the moving-staircase compared with the lift attests to this. It is not merely speed, but the possibility the moving-staircase gives for independent individual action as opposed to collective action dependent upon an attendant, that is significant. Servants tend to bind and circumscribe action, for their presence makes inevitable the establishment of a routine that only too often rebounds upon their employers.

Self-service has the merit of engendering responsibility and of enhancing awareness as well as of increasing freedom of action. As unhampered in the Centre as in their own houses, the members are free to improvise to suit all occasions as they arise. As the embryo newly lodged in the womb begins to build its cells into the substance of the uterine wall, so each new family emboldened to strike out for itself in this living social medium can add its own quota of 'organisation' to the Centre—the outstanding characteristic of which is the abiding fluidity of its constitution, permitting continuous growth and the functional evolution of its society from day to day and from year to year.

So in the Centre there are no attendants, no waitresses. This means that where possible all equipment has had to be designed to be handled by the members themselves. In the main the furnishings are light stackable tables and chairs which can be moved from place to place as occasion demands; the cafeteria utensils also are stackable and devised to be taken and replaced by the members. These are seeming trifles, but they have their far-reaching significance in the type of social organisation that is growing up in the building.[1]

In designing the building it was far from a simple matter to assemble this type of equipment. It is often not recognised that though quick-service installations exist here and there in restaurants, there has been no *self*-service in this country. Much research and ingenuity therefore remain to be devoted to this subject to bring it to anything like practicability. The initial stages moreover are costly. To give a simple example: £60 had to be paid for the first mould for a simple article such as an unbreakable plastic plate-saucer; in use it was found to chip readily and had to be replaced by a metal one, for the initial pattern of which a further similar sum had to be paid. A suitable design once on the market, no subsequent Centre would have to incur these initial charges. This instance is only one of the innumerable small difficulties encountered in equipping the Pioneer Health Centre. Designs for durable and accurately stacking bowls and glasses had to be worked out and produced.

[1] See photograph 14, p. 57.

Again, there is no apparatus made for the orderly reception of dirty utensils in soaking-fluid from which they can be removed and washed the following morning—thus avoiding unnecessary night work. Modern materials should make all these possible and cheap, but they still remain to be designed, made and reproduced.

The provision of facilities for self-service in all Centre activities, means that the cost of maintenance and of staff salaries remains more or less constant. The cost of running the Centre, therefore, need not rise with increasing membership, though obviously as new activities of the members arise, certain further capital outlay is necessary.

The Key.

Another instance of the necessity for designing not only a new type of building but also new installations to meet its purposes can be given. Research requirements necessitate a record of entrances and exits and the times of day when these occur, as well as a record of the activities of each individual once in the building. It is also very desirable to devise some impersonal method of collecting subscriptions, and to minimise the cost of collection. These requirements have led to the formidable task of the invention of a multiple purpose 'key' for the use of each individual of a member-family. This key is a small circular disk $1\frac{1}{4}$ inches in diameter and $\frac{3}{8}$ inch thick, with a ring to hang on the watch chain or key ring. It gives access to the building to each individual of every member-family and forms an entrance tally recording on a moving sheet the family number, which individual of the family, and the time he enters; it passes him in for a week after the family subscription has been paid, but excludes him and his family until this duty has been fulfilled on the 8th day; it also permits of his attention being called on entering the building if for any reason he is wanted. Besides this, by means of subsidiary recording machines, it affords entrance to any special activities on payment of the required sum, again giving a record of entrance and time for statistical purposes. Suppose the scientist should wish to know what individuals are using the swimming bath or consuming milk, the records made by the use of the key give him this information.

It was three years before time could be found to design and

construct this instrument.[1] In the meantime subscriptions and payments have had to be taken and recordings made by individuals whose time has been largely devoted to this purpose only. But the organisation was from the beginning planned with an installation of this nature in view; it is as necessary for the purposes of research and for dis-embarrassing the observer of duties connected with control, as it is desirable for freeing the individual member and useful in diminishing the running costs of the Club.

Attitude of the Staff.

How does all this affect the work of the biologist? As observer moving about the building, he now comes to be in much the same position as any member—able to exercise his preference in action. As we have seen, these preferences are for the accumulation of knowledge and information about the families gathered there. The branches of knowledge in which he is particularly interested are, as we have also seen, those which concern the biological capacities of the families, and those which concern the actional or social capacities of the individuals of those families.

Upstairs in the physiological department, medically trained biologists have a group of instruments which the families use, thereby giving a record of their physiological balance. Here then is our biologist, appearing primarily to be occupied with the task of periodic health overhaul set in the family environment in a guise sufficiently like that of a family doctor to make him familiar and acceptable to the family. But, it may be asked, is he acting in a medical capacity?—and that is a most important question. He needs close watching. Perhaps before an answer can be given, the aims and practice of Medicine will need redefinition in the terms of modern Science. All we know so far is that the biologist is among families to find out facts, to disentangle reality from illusion, from phantasy, even from aspiration—whether his or theirs. He is there to look at each family; not as it could be, but as it is and as it will act in the new conditions created.

But there is another aspect of this biologist's work. On the social floor of the building there are also 'instruments'; for all

[1] The key was designed by Dr. G. Scott Williamson and constructed under his direction. It was paid for out of a bequest for research from the late Sir Henry Van den Bergh and was about to be installed when war broke out.

the equipment throughout the building—the billiard tables, the skates, even the floor space—from the biological point of view, *are instruments for the development of the potentialities of the member-families*, while at the same time being the means for assessing the growing capacities of individuals and families alike. Of these instruments for action and recording, the social biologist moving constantly throughout the building is Curator, keeping the use of everything available for those desiring to use it, and at the same time recording the manner and results of its use.

The training of the staff is difficult. It is in fact no easy thing for the individual as a scientist to place himself as an instrument of knowledge completely at the disposal of any and every member, and at the same time, without exercising authority, to assume his right and proper position in the community as a social entity. But he is also there to make observations. This the members have readily come to accept, jokingly describing themselves as the biologist's "rats."

They soon come to appreciate that the scientist's primary concern is *to be used by the members as a means of reaching and sustaining their own maximum capacity for health*. Moreover, they come to sense that in carrying on their own activities and inaugurating new ones through the method of self-service, many of them are in fact step by step themselves growing into important members of the staff.

So at the end of four years there is little to distinguish members from staff in the social interplay of the Centre. The whole medium is social—Science socialised. The Centre has, in fact, shown itself to be a potent mechanism for the "democratisation" of knowledge and of action.

CHAPTER V

HEALTH OVERHAUL

We have dealt with the necessity and the nature of manifold 'chances' in the environment for the study of family function. The mention of swimming bath, gymnasium, dance floor demanding large and obvious areas in a necessarily extensive building, might make it appear that those particular 'chances' for physical and social action took first place in the minds of the organisers This is far from the fact.

Physical co-ordination is the first and obvious step by which the faculties can be built up, thus equipment for physical activity is there because it is imperative to begin on as primary and simple a basis as possible. But this does not mean that those responsible for planning the Centre are sensible only of physical development.

Fundamental to the design of the experiment and pivotal to the organisation, other and still more important material is assembled in the building for the continuous and spontaneous use of its members. As prominent as the centrally placed swimming pool is the 'pool of information' into which members from the first moment of joining are invited to dip.

This pool of information is primarily located in the physiological department, popularly called the "medical department", where the family overhauls take place. But as the Centre has grown and as staff and members alike have come through practice with a new instrument to know more of its use and possibilities, it has become clear that *all action* in the building is illuminated by knowledge from this source. Facts which the member-family first meets with in the physiological department are continuously being digested through experience in action throughout the building, added to through contact with the staff in all other departments and confirmed through association with other member-families grown familiar with their meaning and with the use of this source of information. Unlike the casual visitor then, those who dip into this 'pool' do not mistake the main drift of what they find in the Centre ; while some in a short four years have come to sense the far-reaching significance of the service.

Periodic health overhaul for every individual in the family is a primary condition of membership, and the first examination takes place immediately on joining. But the family do not come to their examination without understanding its purpose and the advantages to be derived from it. Nor, indeed, do they come without first having had the opportunity of meeting the 'doctors' who are to examine them.

What is their first glimpse of this new opportunity ? Usually, before joining, they come to look round the Centre. Sometimes they are brought by friends who are already members. If not, they are introduced by one of the staff to some Centre-member who takes them on a conducted tour of the building. During the course of this tour the physiological department is visited, and the family is introduced to the receptionist. Possibly during this visit various members—men and women—are seen flitting through the department in their conspicuous and attractive examination gowns. Their ease of movement and the absence of apprehension on their faces does not escape the notice of the visitors. Perhaps a family is emerging from one of the consulting rooms where a 'family consultation' has just taken place ; perhaps the last few words between the doctors and the family are being said in the passage. Obviously, relations between them are of an easy nature ; there is nothing formidable about what goes on in this department. So, fears and alarms, that cling to the anticipation of a "medical" in the minds of many, unconsciously meet their first solvent in this informal tour of inspection.

Enrolment.

The new family think they would like to join, so on leaving the reception room they make an appointment to meet the 'Doctors' for an *Enrolment Talk.* "You will hear from them better than I can tell you what the Centre means", says their guide, as she shakes hands and leaves them with the receptionist.

The day comes for the Enrolment Talk. It is probably a Saturday they have chosen, because the father is free and the mother will have the children clean and tidy for the Saturday evening shopping expedition. Together they come into the consulting room and meet the two 'doctors' who later will examine them, and the biologist who is Curator of equipment. Introductions follow and all sit down. After a little preliminary talk, from which the biologists derive a hint as to the disposition of the family as a whole, the talk may proceed :—"The Centre is here for the use of families like yourselves at a time when it can be of the greatest value to you—that is to say, when you are young and the children are growing up. Everyone wants a full measure of health and a chance of doing all that he is capable of. Perhaps, therefore, the most important thing the Centre offers to its members is the chance of finding out something of what their own capabilities are. So you will find the health overhaul of the family which takes place about once a year, or every eighteen months, of great use in helping you to know where you stand in this respect. The way it is done is for each of you in turn, beginning with the father as head of the family, to come to a laboratory examination.[1] After there has been time to complete the results of the laboratory tests, you then come, one by one, to one or other of us (indicating the man or woman doctor present) for your personal overhaul. When this is all finished you make one further appointment for what we call a '*Family Consultation*'. At this consultation we all meet again here. We then tell you what has been found and—as far as we ourselves know—the meaning of these findings. If anything is wrong, we shall be able to let you know if anything can be done. That will leave you in a position to put it right, if you wish to do so ; or, if any question or difficulty arises, through its discussion a solution may be found for it."

"Meanwhile, any information *we* may gain during the examination or consultation will be of great use to *us* because our interest is to study health and find out how best it can be achieved. If our members want to know anything, or are worried about anything, they always come up at any time and ask for an appointment to see either of us. So you must never be afraid that

[1] Besides being head of the family, a position to which formal recognition is due, the father is also the member of the family most likely in the first instance to fight shy of examination. It is useful, therefore, to make the family overhaul contingent upon his initial compliance. The disinclination to be examined does not result from any lack of courage, but appears to be derived from his or his fellow workers' experience of medical overhaul in connection with his work, where to the worker the results seem to operate, as often as not, against his general interests rather than for him. But, the first family consultation once over, it is often the father who makes the most use of the opportunities that the health overhaul of the Centre affords the family.

any matter is too trivial to be of interest or importance.[1] It is the little things that matter; a chance taken at the right moment, 'a stitch in time that saves nine'—perhaps in medical matters ninety-nine!"

They all go out. Only after this talk do the family decide whether they will join the Centre or not. If they decide to join, they there and then sit down in the reception room, become enrolled as a member-family, pay their first weekly subscription and proceed to make their individual appointments for laboratory overhaul in the sequence which the family situation demands. If the mother is pregnant, or if there is a young baby, the mother and baby will make the first appointment instead of the father, for at this period they represent the 'growing point' of the family and so dominate the situation for family and biologist alike.

Few people not acquainted with the circumstances of the weekly wage-earner realise the significance to him of this method of approach by choice and at his own time. He expects to have to fit himself into other people's convenience in all that he does. His hours of work are inexorably determined for him, he is not free to arrive an hour earlier and leave an hour earlier as is his master; if he wants a job, or hospital treatment, he must attend at some hour determined by those who have either at their command. If his wife wants to see the child's teacher or the school doctor, she must attend at some stated time not necessarily convenient to her. Nowhere in fact—except by the private dentist or the hairdresser—is the time and convenience of the weekly wage-earner and his family considered as anything to be respected. Perhaps it is this background rather than any other factor (e.g., ill-will or ignorance on the part of the applicant) which has led to failure of the attempts on the part of the hospitals to introduce an appointment system. Besides the hospital being situated at a relatively great distance from the home, the times available are insufficiently elastic to meet the family's circumstances. It is useless to hope for appointments to be kept if the only hours offered are ones which clash with other responsibilities the individual recognises.

The truth of this seems to be borne out by the fact that in

[1] The number of families or individuals who have abused this invitation has been very small.

the Centre it is the exception for appointments not to be kept, and if an individual is prevented from coming it is quite unusual for a note or message of apology not to be brought by some member of his family. The reasons are obvious: the Centre is situated within the range of the daily excursion of its members; it is organised primarily for their convenience; the staff are known personally to them, and there is a continuity anticipated in the mutual relationship between the Centre and all its members. The first thing the new family notice, then, is that the Centre does seem to belong to and really is organised for its member-families. That in itself is arresting and to some of its members at first even difficult to believe. In some cases it has happened that the father has hesitated to give his whole-hearted allegiance to the Centre till he has assured himself that there is no 'snag' behind so unusual a situation.

The Health Overhaul.

The purpose of *health* overhaul is not the discovery of sickness in the members of a family—that is merely incidental. The health overhaul is an attempt to estimate the physical efficiency of the family and all its members, and their capability for:—
(i) individual life; (ii) family life; (iii) social life.

What shall we look at first? The body of each individual: build, type, soundness or otherwise of the component parts, the nature of the effluents and the body's economy. A bench test, in fact, is done of the working machine of each individual. This involves a laboratory examination[1] and a complete personal overhaul, which has already been described in detail in "Biologists in Search of Material".[2]

There is no hurry. No presence of, or fear of sickness impels urgency, so that the examination can be carried out in the family's leisure. The examination department works from 2 till 10 p.m. five days a week. This wide range of hours enables each to choose a time to suit his or her work and convenience. Each goes first to the laboratory for collection of specimens and other laboratory tests by a bio-chemist. Nearly every one, but particularly the men and the children, enter into this part of the proceedings with lively interest.[3]

[1] See Appendix iv.
[2] p. 43-51.
[3] See photograph 5, p. 54.

On a subsequent day, also at the member's convenience and by appointment, comes the personal overhaul in the consulting room conducted by the biologist, a man for the men, a woman for the women and all children under seven. For this each individual is asked to strip and is provided with a suitable and clean examination gown.

The examination wraps made of artificial silk are sterilised between each using by the simple procedure of being run through a domestic dry-cleaning machine. The members appreciate the seemliness of this detail of organisation. On two occasions only in four years has any member appeared with even a vest under the examination gown.

The Family Consultation.

When all have been examined and all facts collected, there follows a final appointment, this time for the *Family Consultation*, in session with the man and woman biologists who have conducted the individual examinations. The procedure is as follows :—When all are seated in the comfort of an informal setting a review is made of the findings collected in the laboratory and at the personal examination of each individual. Beginning with the youngest child, the results of each laboratory test are given, followed by the results of personal overhaul. The significance of these findings wherever possible is translated into lay terms for the family, the object being that they should themselves understand their meaning. It is the family as a whole which receives this information.[1]

The youngest child finished, it goes out and the next child comes under review. So the consultation proceeds till all the children are gone and the mother and father alone remain with the biologists.

By this time the parents' interest will have heightened and usually they will have drawn closely together. A difficult partner may already have begun to thaw out, for after so much information has been forthcoming there is much to discuss about the children,—the meaning of the findings for each child, for the other children, and for the family as a whole. There are now questions the parents wish to ask, and remarks the biologists wish to make in the children's absence. Over the discussion of these subjects,

[1] See photograph 6, p. 54.

now of mutual interest to family and biologists alike, confidence grows, so that by the time the husband's turn comes any misgiving or shyness he may have had about the review of his own overhaul has vanished. The findings of his examination now follow in the presence of his wife. He is deeply interested. Last of all the review of the wife's overhaul with its lay interpretation completes the family picture. She too seems somehow to gain in dignity through this discussion; her place and importance to the family is confirmed by the procedure.

Then follow any questions and comments that either doctor or member has to make. By this time, foregone intentions, almost forgotten hopes lying deep in the family consciousness have risen to the surface and somehow seem to take on a new potency in the face of the facts that have led to their unfolding. The meaning of the Centre with its exceptional chances for the whole family dawns clearer to both parents as they see it bringing their aspirations once more into the realm of practical possibility. This is often a very moving moment.

Also there may have come to the surface long buried fears, anxieties, dread unspoken yet ever active in the minds and lives of one or both parents. All these anxieties, too, take on a new aspect in frank discussion of the facts that led to their uncovering. A sense of relief follows either the possibility of doing something about them, or facing the inevitable with full knowledge as well as with the assurance that nothing undisclosed now lurks in the family situation. Example after example of these two types,—unfulfilled aspiration, or secret and unexpressed anxiety, could be given were they not to overburden the text. The effect of the family consultation for the parents is like that of watering the parched soil in which a wilting plant has been struggling to grow!

Through this medium of a family consultation following the periodic health overhaul, the biologist and the family looking at the facts together find that they have been moving towards some common basis of language and understanding. Experience, facts and knowledge shared—this is the beginning of a relationship that is to grow and to deepen in the continuous use of the club and of all its activities. The parents begin to understand some seemingly unusual features of the service that have struck them as peculiar. For instance, they may catch the sense of health as being not the mere absence of disease, but that which concerns their action and their growth; not something which can be 'given' but something they, themselves, must *take*. They may comment upon the fact that there are no rules as to what they shall do in the Club, and begin to understand why, in an organisation concerned with health, they are left free to choose and to act for themselves; why there are no professional organisers or instructors as there are in holiday camps and most other clubs;—only the materials with which to plan things for themselves and build up their own activities. To others it becomes clear why all the facilities of the Club, though each subject to small payments by the adults over and above the family subscription, are nevertheless free to all children. They are the members of the family for whose growth and development the need for equipment of all kinds is greatest.

But the family consultation is not yet over. If the parents are old and their children grown up and married, much time is usually spent in hearing of their doings and of the parents' hopes and anxieties for them. But if the parents are still young, and if they have young children, some insight is given into the significance of parenthood for them all. In this case the consultation will go forward somewhat on these lines.

"From many different sources Science is beginning to piece together facts that have a bearing upon parenthood. Some of these will interest you. At marriage you both entered on a new stage of development, in which instead of going along on your own individual ways you joined forces and became one—a unity. From the moment of courting you began to act on each other, changing yourselves and changing each other. This was not merely in little ways of give and take, in which you adjusted your tastes and habits; it goes much deeper than that, for changes actually began to occur in your bodies, so that the physical constitution of each of you was modified and you became more and more attuned to each other. The influences which affect such changes are so potent that they do not even need close physical contact to set them off, and when development goes smoothly they are so natural—like many other subtle though profound influences which affect our bodies in the course of life—that they are often taken for granted and overlooked. Pollen of grasses is a good example of how strong but subtle the things that affect us can be. A healthy individual enjoys walking through a hay field, not realising that he is in fact 'digesting' what may be a valuable meal of pollen. It is only when function is impaired, when something is wrong and the person cannot 'digest' that pollen, that he rejects the experience and gets hay fever. Only then does he *become aware of the high potency of the pollen*".

"These changes that take place through proximity and even through casual contacts, are, of course, deepened with the intimacy of marriage. Indeed, some people change so much at this time that, for instance, they find themselves making an entirely new set of friends ; their old friends do not seem 'quite the same'. But if that was your experience you will probably see, if you look carefully, that what really happened was that *you* changed ; not your previous friends".

"It is important, therefore, that when folk are courting and when they are newly married they should have as many channels as possible open to them for making new friends to meet the measure of their development and growth at this time. That is one of the reasons for the way the Centre is planned, with its chances for meeting many people and for families to do things together. Here a young family can mix with some hundreds of other families. Out of these they will inevitably make many acquaintances, from among whom they will be able mutually and naturally to select one or two intimate friends whose friendship as it deepens will enrich the home husband and wife are building together".

"These profound changes that we have been speaking of, which begin in courtship and early marriage, reach their height in pregnancy. During this period the woman's body changes its constitution. Her tissues become fluid and softened so that, for instance, in some you can almost bend the bones under the finger. All the tissues undergo this change,—liver, bones, hair, eyes, brain—all are flooded with the circulatory fluids of the body so that their essence seeps out and is carried round in the blood to the womb where the growing child can draw upon it for its nourishment and growth. So the child is made not out of the food that the woman eats but out of the very essence of her body. It is tinctured with its mother's own 'specificity'—as we call it".

"This process is carried still further through lactation. Only after this is over does her body as it were 'harden' again, but now into a different form for she has become a *mature* woman. By and through what has happened she herself has developed and has changed".

"But the strangest part is that by this process which has gone on in her body during pregnancy she has changed the man too. We do not yet know how this is brought about in human beings, but we already know much about it in some of the lower animals". At this juncture, according to the comprehension and interest the family evince, the point is illustrated by some example found in nature, such as that of the nesting doves given earlier in this book.[1]

"So it makes a real difference to both husband and wife that they should be together while the wife is carrying and feeding her baby. Probably the old wives were not so far wrong when they said the milk could be 'soured' by untoward events and disruption of the home during these important and subtle periods of development for the pair. It is in the light of such facts being gathered by the scientist that we begin to see that the child is not just a delightful and special sort of luxury which a young couple may or may not be able to afford. It is Nature's means by which both man and woman *reach their own maturity*. In healthy parents each successive pregnancy will hold some possibility for changing and maturing both of them".

"It is true that increasing development and maturity does not always result from pregnancy. That does not necessarily mean that the pair are not capable of such change but that the necessary circumstances for the changes to occur, are for some reason or other not favourable".

"Now it is clear that if pregnancy is so important a time in family development, both parents must be in their fullest health when conception occurs. Only in this way can they get the best out of it and through it reach a fuller measure of health. Indeed, any physical weakness or deficiency, perhaps not noticeable at other times, can rise into prominence during the great activity of pregnancy, and the pregnancy thus lead to an *undoing* rather than a building up of the life of the whole family. So, in order to be sure that you are as fit as you can be when conception occurs, it is very necessary to put everything right before-

[1] See Chap. II, p. 37-39.

hand, and to make up any deficiencies that you may know of. This sometimes takes a little time. You may therefore find that you need advice on the control of conception. There are very few methods that are relatively safe and not harmful, and if you do need any help or advice on this subject you have merely to ask the receptionist for a 'special appointment'. The matter can then be discussed with one or other of the doctors, who will advise you as the circumstances demand". In this way, advice on the control of conception is available to all parents on joining the Centre, taking its place as *a measure related to the health of the family.*

The consultation ends by letting the family know that provision is also made for pregnancy. "As pregnancy is such an important time for the parents and for the coming child, the Centre also makes full provision for the maintenance of the health of the expectant mother from the time of conception throughout the whole pregnancy. Special arrangements for delivery are also available for all families who already enjoy a full measure of health. . . .'

The interest with which this talk is received is one of the most encouraging features of our method of approach to the family. The impersonality of the situation has already been achieved by the objective discussion of the findings of the overhauls ; shyness has evaporated, and reticence been swept away. It is one of the most usual occurrences for one or other, if not both partners again to contribute some personal experience in support of what they have just heard. For instance one will remark—"Perhaps that explains why old married people so often come to grow alike", or—"That happened to us with our friends and quite worried us because we thought we were fickle, and that it was in some way our fault that we lost sight of them when we married!" Or again the man may instance some ache or pain that he always suffers during the pregnancy, one having stated that he knew which day the confinement was going to take place because though not given to vomiting he always had an attack on that day! We do not give these examples as *evidence* of the facts in question, but only as an illustration of the interest with which the subject arouses and the quality of comprehension with which it is met.

The family consultation, then, is a means of giving to the

family and all its individuals the facts that can be gathered from the health overhaul, and of passing such facts through the family mill for digestion. This form of consultation affords a mechanism open to the use of the biologist at all the crucial phases of family development; for instance, after the announcement of conception, after the birth of the baby, at the time for each weaning as it occurs—breast weaning, skirt-weaning of the child from dependence to the state of independent mobility—and so on right up to the time of puberty when the child begins to extend the home circle far into the greater world. The family consultation is a mechanism available also at the behest of the family at any time a difficulty or question may arise in their minds. They are free to ask for one whenever they wish to do so.

Once the first family consultation is over, families find no difficulty about a second dip into this 'pool of information'. They come with a certain anticipation to subsequent overhauls, each of which seems to pick up where the last left off, and goes on to embrace the new circumstances that in the interim have arisen in the family and the new developments in its individual members. As the children grow up the circumstances of their schooling are discussed; or, well in advance, the choice of a career is considered so that the family can prepare for any financial call that might, if left to the last minute, stand in the way of the child's chances. The biologist being in possession of knowledge of the constitution, tendencies, and often capabilities of each child and of both parents through the periodic health overhaul and from continuous observation of their actions in the Centre, is in a strong position to help the parents to make a wise decision as to future careers for their children. Discussion at the family consultation often helps to dissolve any prejudice which might lead either parent through ignorance to impede the child's development.

That all the examinations and consultations, following the traditions of clinical medicine, are kept strictly personal and confidential and are carried on without any sense of hurry goes without saying. Trouble is taken from the first moment of joining to engender between staff and members all the amenities of social contact, and by prompt and careful attention to all details to make it clear that the Centre is there primarily for the use of the members. This attitude is quickly appreciated, and almost universally understood by the members and responded to by confidence placed in the staff and workers. Without this lively response the Centre would lose much of its value to members and to biologist alike.

It must be remembered that more than half the difficulty of establishing contact with people moving in different circles from ones own, lies in the absence of a common basis of experience. This difficulty is naturally greater where those approaching each other are separated linguistically by forms of expression as divergent, for example, as that of the scientific worker accustomed to speaking in technical terms, and that of the man in the street who 'feels' his way through life and has comparatively little need to express himself in any but intuitive gesture. Mutual action leads to, and must take place before there can be mutual confidence and understanding. But, equally important, unless that mutual action has concurrently presented the less articulate participants with language with which to speak of their common experience, the absence of language will still bar fruitful communication.

The choice of our particular populace has been a happy one in that the members are characterised by a directness and simplicity that enables them, as the experiment proceeds, to pick up language to describe what is within their experience, while remaining quite unresponsive to ideas outside their own ambit. There is therefore no danger in this setting, as there would be among intellectuals, of their adopting ideas which have no counterpart either in their experience or in their capacity to experience. This is a point of very great importance, where observations are to be made, an educational programme is envisaged or a technique of homoculture is contemplated.

The very first family consultation then, gives the family a point of reference in words and facts which will grow in value with the recurring contacts arising out of continuous membership with its regularly repeated overhauls and consultations, and continuous association both with staff and members in the use of the club. Each or any of these various contacts in turn may prove to be a peg the observer finds upon which to hang still other facts as they become pertinent.

From what has been said the reader will begin to see how, from this pool of information in the physiological department, a continuous trickle of pertinent facts flows throughout the whole Centre organisation, tending—like any fluid—to spread gradually over the whole field. In doing so it finds its own level in each situation where it lodges. In this way member-families are always spontaneously gathering knowledge through their associations in the Centre. Thus, *facts acquire their meaning in virtue of their topicality*; available at all times in the physiological department and demonstrated in action in the life of the Centre and its members, they rise into prominence for each family *as it acquires the ability to act upon them.* The source of information is always there, but, like the swimming bath, it increases in significance with the use that each family can make of it.

Most important of all, the family consultation illustrates the fundamental technique which the biologist must follow in all contacts, whether it be in the presentation of facts, or in the provision of instruments of any kind in the immediate environment. The presentation must not be directed to one or other individual, but *to the family as a whole*. Here in the family consultation we have a setting in which this can take place with naturalness and suitability.[1]

It will be seen from this commentary that the biologist does indeed attempt, and in a great measure achieve, the position postulated as necessary for the experiment, namely that of being in mutual relationship with the members—giving back to them in terms of facts and knowledge an equivalent for what he takes from them for his own studies. Only if, through mutual action in a topical approach to each family he can induce a process of continuous development, can he hope to gather repeated harvests in his chosen field of study—that of human function.

[1] When the children of a family are grown up the results of their overhauls are discussed with them individually at the end of their own overhaul, and they are not required to attend the family consultation unless they wish to do so.

CHAPTER VI

FINDINGS OF OVERHAUL

ALTHOUGH this book is concerned with the study of *Health* and with the experimental investigation of the technique required for its cultivation—a procedure distinct in principle, aim and methods from the study and practice of Sickness—it would be incomplete without reference to results of a clinical nature that have emerged from the use of these new methods. Periodic health overhaul of member-families has yielded ascertainable facts in the clinical field about which a definite statement can be made. When dealing with such material we pass into the realm of pathology, so that it becomes necessary to set aside for the moment the functional considerations already put forward, and to consider the contents of the first two sections of this chapter mainly from the point of view of traditional Medicine.

In 1938, under the title of "Biologists in Search of Material", we published a review of the first 500 families overhauled.[1] Since that date we have had a further 2¼ years' experience in the use of the same technique on the same and on similar material; i.e. on families who have continued as members from the time of the first report and on others joining later. A review of this material has now been made by the senior members of the Staff working together in close co-operation since the outbreak of war.

This second review was of 1,206 families, comprising 4,002 individuals of all ages. The overhauls were conducted according to the methods described in detail in "Biologists in Search of Material" and amplified in this book by a description of the various 'family consultations' the technique of which has been evolving gradually as our experience has grown.

In 4½ years, of the 1,206 families examined, 877 families had one overhaul, 227 had two overhauls, 96 had three overhauls, and 6 had four overhauls, at intervals of a year to eighteen months. The overhaul in each case was a complete one, including laboratory investigation and personal examination of each member of the family, followed by a family consultation. As part of the routine of examination, all infants were seen weekly till taking table-food and later fortnightly until walking; monthly

[1] See Sect. III. pts 1 and 2. p. 52 et seq.

till between 3 and 4 years old, when the interval extended to three-monthly. During school years the children were usually seen at six-monthly intervals, but it must be recalled that all children using the Centre came within daily observation of the staff. The special care of the family during the course of pregnancy is described in detail in chapter VIII.

Besides this routine procedure, between the regular overhauls of each family these same individuals were also examined:—

(1) on discharge from medical care after any intercurrent illness;

(2) at their own request if suspecting anything wrong;

(3) at our request on noticing any abnormality in any member whilst in the club.

In addition, many had extra family consultations and special 'parental' consultations to be described later in this book.

The pathological conditions found in the course of examination, and their incidence among the populace, run so closely parallel in our first and second reviews, and the second confirms so closely the findings of the first review published in "Biologists in Search of Material", that we do not propose to burden the reader again with a full account and discussion of them, but shall proceed here to discuss only certain aspects of these findings which have a bearing upon the question of medical administration in general.

The first and outstanding finding is that from a total of 3911 individuals of all ages,[1] 3553 (90.85%) at first overhaul were found to have something the matter with them, i.e. some physiological defect, deficiency or aberration,[2] As the district from which these families were drawn was chosen because it did not contain a social-problem group of the populace, but on the contrary one that was considered likely to yield a relatively healthy populace, this finding is an arresting one. It indicates that the field in which modern medical science could be applied with benefit is far greater than that at present visualised by the most advanced advocates of a rational medical service.

In view of the indications from many other sources of a greater prevalence of disorder than is usually recognised, this finding

[1] This figure excludes the 91 infants born within the membership of their parents, and on whom the Centre therefore had its influence before birth.

[2] For the age and sex distribution of these disorders in our populace, see Chart I (*frontispiece*).

cannot be disregarded on the score of being a solitary and unique survey of its kind.[1] In 1941 among the first batch of American recruits, 50% were rejected as being unfit for admission to the U.S. Army, and in the opinion of the authorities it was unlikely that more than 10% of the rejects could be made fit for service.[2] This indicates that the disorders found were not of a merely transitory nature, and leads to the conclusion that our findings are not peculiar to Peckham, nor even to the British Isles. It is a general, not a local phenomenon that we have encountered. The interest of the findings on the U.S.A. Army recruits is that the 50% of rejects were all young men ; that is to say they were of an age when the health of the individual is usually regarded as likely to reach a relatively high level. Amongst our own populace the incidence of disorder in males at this age as compared with other ages can be seen from chart IA (*frontispiece*).

Let us examine more closely this 90% of our members with something wrong. Approached from the clinical point of view there is nothing remarkable about the disorders found. In spite of the fact that these individuals were going about their daily work their disorders are just those listed in any text-book of Medicine, the defects ranging from the most trivial to the most serious condition. It is then not the *seriousness* of his disorder that immobilises the individual, nor, as we shall see, that converts him into a 'patient'.

Approaching our members through *health* overhaul, we were faced not with 'patients' but with the man in the street. Our experience therefore differed widely from that of the clinician. It was for this reason that as well as classifying the disorders from the clinical standpoint which is an *objective* one,[3] we were led to make a second classification according to the reaction of the individual himself to his condition ; i.e. a *subjective* one.[4] This second method of classification has presented us with medical problems in an entirely new and interesting light.

Classified from the subjective point of view, the individuals *over 5 years of age*[5] examined by us fall into three categories :—

[1] See for example *Medical Examination of* 1592 *Workers*, Morris, J. N., *Lancet*, Jan. 11, 41, p. 51.

[2] Final Report of Temporary National P.C. Committee. U.S.A. 77th Congress. 1st Series. Senate Document No. 3589.

[3] *Biologists in Search of Material*, Section III, Part I, p. 53-76.

[4] Ibid. Section III, Part II, p. 77-91.

[5] It is not possible to arrive at any subjective classification of individuals under 5 years of age.

(1) Those in whom disorder is accompanied by dis-*ease.*

(2) Those in whom disorder is associated with a sense of *well-being.*[1]

(3) Those without any signs of disorder (? the healthy).

If, of the total of 1,206 families examined in this our second review, 500 families are taken at random for comparison with the 500 families examined in "Biologists in Search of Material", we find in individuals of 5 years of age and over, the relative proportion in disease, in well-being, or without disorder to be very much the same.

	1st 500 *families.*	*2nd* 500 *families.*[2]
Disease	31.6%	21.3%
Well-being	59.0%	68.5%
Without disorder	9.4%	10.2%

1 DISEASE

In considering this category, we cannot do better than quote from our previous Report :—"The word 'disease' is used very loosely, not only by the public but also in most medical text-books. 'Disease' may mean (1) the subjective state of the patient, (2) the objective findings of the professional diagnostician, or (3) both the objective findings and the subjective state as though they were either interchangeable or one and the same thing. *To avoid this confusion here we shall use the word 'dis-ease' to mean only the subjective state of the sufferer. The objective facts discovered by the professional diagnostician we shall term 'dis-order'.*"[3] To this definition we shall adhere strictly throughout this book.

The condition of disease, then, arises from a disorder or disorders that have led to pain, to discomfort, to disability, or to limitation of action in the occupational, family or social life of the

[1] Ibid : p. 78 and 83-89.

[2] The relative numbers are as follows :—

	1st survey	*2nd survey*
Disease	484	328
Well-being	902	1052
Without disorder ..	144	156
	1530	1536

For an analysis of the age and sex distribution of disease, well-being and those without disorder in the second survey, see chart II (p. 98 and 99).

[3] *Biologists in Search of Material,* p. 78.

individual. One or other of these states is always present before the sufferer becomes aware of his 'disease'. Though all disorder is not accompanied by disease, all disease has, of course, its underlying disorder. The sufferer's appreciation of his disease, however, may give him no indication of the nature, nor gauge of the severity of the disorder. He may suffer what to him is serious disease from a trivial disorder, e.g. nettle rash, fissure in ano ; or the slightest symptom from the most serious disorder such as cancer.

What then is the value of the distinction we make between these two states, dis-*ease* and dis-*order* ? Its importance lies in the fact that it is upon the degree of the patient's disease that hangs his decision to take action for the removal of his disorder. It is *disease*, in fact, that converts the individual into a 'patient' ; not disorder. The patient gives little or no consideration to the disorder underlying his disease ; it is disability and interference with his actions that leads him to recognise that something is amiss, and that consequently leads him to decide that treatment is necessary. So under the existing regime, it is inevitable that the *ignorant patient* is the primary diagnostician of the existence of his sickness.

The medical services of the Nation (with the possible exception of the School Medical Service and the Child Welfare and Maternity Services[1]) are designed for this 'patient' : that is to say for those in the subjective category of disease. That the doctor has no means of coming into contact with the man in the street until by seeking advice the man constitutes himself a patient, makes this clear.

In our two surveys, we find some 25% of individuals in disease. In the face of the great advance in medical science, we must presume that these individuals in disease could derive benefit from skilled diagnosis and treatment, and, in view of the ever increasing facilities for treatment, it would be reasonable to suppose that they were doing so. In fact, however, less than one-half of those in the category of disease were receiving medical care.[2] The remainder, though conscious of disease, were

[1] It is claimed by us that both infant and expectant mother are in fact *converted into* 'patients' by the circumstances of the existing provisions made for their Welfare : e.g. congregation in Chinics and Hospitals with a clinical atmosphere and approach.

[2] In the survey of the 1st 500 families, of 484 individuals in disease, 121 only were in receipt of medical treatment ; in the second 500 families, of 328 individuals in disease 124 were under medical care.

CHART IIA

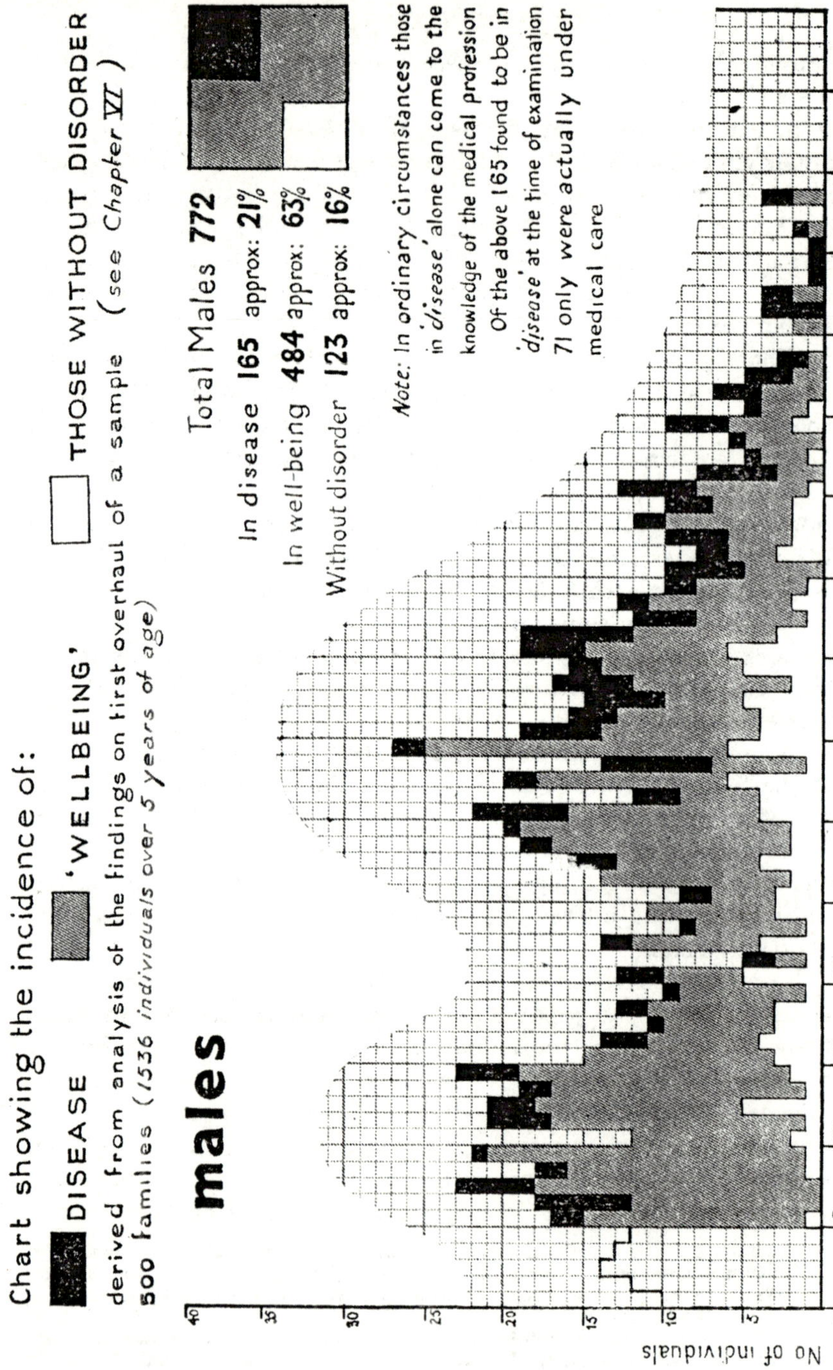

Chart showing the incidence of:
DISEASE
'WELLBEING'
THOSE WITHOUT DISORDER
derived from analysis of the findings on first overhaul of a sample (see Chapter VI)
500 families (1536 individuals over 5 years of age)
males
Total Males 772
In disease 165 approx: 21%
In well-being 484 approx: 63%
Without disorder 123 approx: 16%
Note: In ordinary circumstances those in 'disease' alone can come to the knowledge of the medical profession
Of the above 165 found to be in 'disease' at the time of examination 71 only were actually under medical care
No of individuals
Age in years

CHART IIb

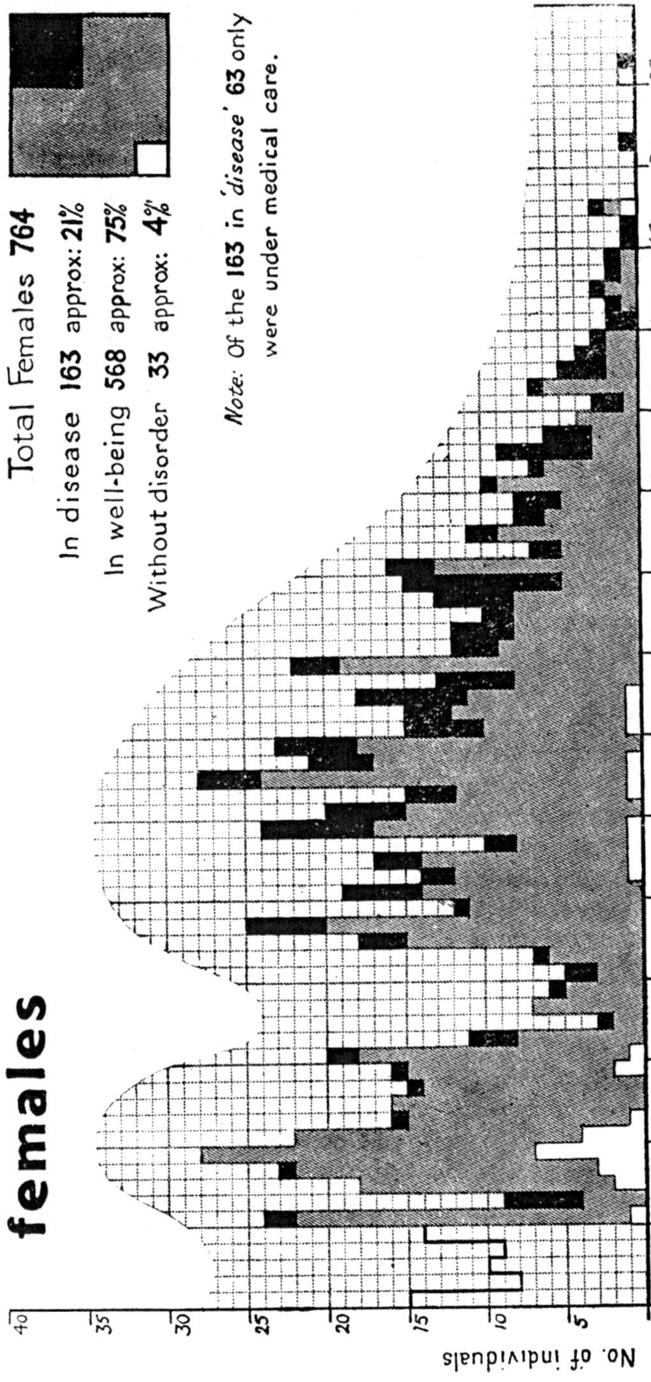

females
Total Females 764
In disease 163 approx: 21%
In well-being 568 approx: 75%
Without disorder 33 approx: 4%
Note: Of the 163 in 'disease' 63 only were under medical care.
No. of individuals
Age in years

not in contact with any diagnostic or therapeutic agency, nor were they receiving any medical attention. Without going into the reasons for this situation, already discussed in "Biologists in Search of Material",[1] let us look at it from the point of view of national sickness and of medical administration in general.

The first point that would seem clear is that what purport to be the statistics of national sickness do not, in fact, represent the *total* incidence even of self-acknowledged sickness in the population, for the statistics are gathered from the number of patients encountered by the medical profession. These national statistics of sickness represent, in fact, only the incidence of medically diagnosed and treated sickness. Where the statistics are being used purely as a measure of the relative incidence of sickness, in one year as compared with another, or one nation as compared with another, this is of course a small point. As long as Medicine is conducted on much the same lines throughout this country and in all other countries, the old methods may well continue to serve this purpose ; but directly any one country begins systematically to apply periodic health overhaul to its families, or even periodic medical overhaul to its individuals, the figures will immediately take on an entirely different aspect. They will no longer be comparable with the old ones nor with those of other countries.

Much more important, however, is the fact that if less than half those conscious of disease are seeking treatment, there must inevitably be occurring a considerable and unnecessary time lag between the recognition of his disease by the patient and the date at which he obtains treatment for it. This is confirmed by the average length of history of symptoms given by the patient when he does finally seek medical advice. So many conditions easily curable in their initial stages are, by delay in obtaining treatment, unwittingly but irrevocably being converted into chronic and incurable conditions, thereby adding quite unnecessarily to the length and extent of the treatment required when the patient does eventually reach the doctor.

The Centre's contact with the ordinary citizen discloses neatly for our attention this unnecessary source of chronicity of sickness. It is chronicity more than any other factor that piles up the cost of the medical services of the Nation ; creates a cumulative

[1] p. 80-83.

loss of working hours and diminution of efficiency in industry; wears down the natural reserves of the individual and continuously robs the family—and therefore the Nation—of its heritage of health.

There is also another side to the question of chronicity—the doctor's outlook. This time-lag militates against the efficient application of modern medical science and thereby prevents the doctor from doing the best work of which he is capable. In the medical profession this is realised to the full. Many efforts have been made to prevail upon the patient to bring his complaints earlier for diagnosis and treatment. But, in the present system of medical organisation, it has not been found possible to do more to facilitate earlier treatment than to create more and better clinics and to distribute them further and wider afield (vide the early treatment of cancer). The effect of these measures is undoubtedly to make it easier for those who *seek* treatment to find it, and that is to be welcomed. It does not, however, touch the main question which disturbs the medical profession, for it does not help the doctor to get into touch with the case as early as *he* could successfully diagnose and treat it.

Although, as we have stated, we found less than half of those suffering from disease to be under medical care at the time of examination, after overhaul in the Centre all, without difficulty, were brought to seek treatment. Thus the Centre's approach —which is not through clinical medicine—is a completely satisfactory one for bringing all individuals suffering from disease to treatment. It is in fact the only efficient method that has been found for the curtailment of chronicity.[1]

Hence we must conclude that although the provisions necessary for dealing with sickness cannot produce health, the organisation necessary for the cultivation of health automatically leads to the rational treatment of *all* those in disease.

2 WELL-BEING

In the second specimen of 500 families, 68.5% of the individuals fell into the category of 'well-being'. All these individuals seemingly well, had something which we could determine as disorder. They themselves however were oblivious of their

[1] For further discussion of this point, see *Biologists in Search of Material*, p. 60-63.

disorders; i.e. free from dis-*ease.* Indeed the majority stoutly asserted that they were in their usual health—"quite well" as they themselves put it.

There is the young childless wife looking physically well, who on examination is found to have a large abdominal tumour. There is the schoolgirl leading an ordinary life of work and play who has an undisclosed nephritis with intermittent attacks of haematuria, overlooked by her mother and escaping notice at the school doctor's intermittent visits. There is the family the members of which as they enter the consulting room all demonstrate an avitaminosis of which they themselves are entirely unaware. There are families every member of which is infested with worms, but where only one child (if any) is known to be so infested and is receiving treatment. This child of course is open to continuous re-infestation from the others who go untreated.

Cancer affords a very good example of a disorder that can be unknown to the individual though easily diagnosable by the skilled physician. During the course of four years, four such cases of early cancer were detected, and as far as our observation carries us were cured; and two pre-cancerous states in women received radical attention from the gynecologist. In four of these cases the individuals concerned were unaware of there being anything wrong; the remaining two complained of some quite trivial disturbance mentioned because they happened to be coming to their periodic health overhaul, but about which they would not otherwise have thought of seeking advice. Perhaps the most interesting of these cases was that of a man who was noticed at the overhaul to be slightly husky. Neither he nor his wife were aware of this insidious change of voice. The doctor was led to pay attention to it because, having overhauled the man at intervals for 12 years,[1] he recognised it as of recent origin. On examination a small papilloma was found on one of the vocal chords. Malignancy was at once suspected and confirmed by section and histological examination at a London Throat Hospital. No glands were yet present; i.e. the growth had not spread beyond the vocal chords. Radium was applied. Eighteen months later, at the outbreak of war—the last date on which he was seen—the man was still completely free of any further signs. Cancer has been taken here as an example because much is made

[1] This was the same man cited in *The Case for Action*, p. 19, 20.

by the public and by the medical profession alike of the 'cancer problem'. The medical profession however is fully aware that all disorders can be equally insidious.

All the instances that we have cited above are those in which the presence of disorder was not recognisable either by the individual or by his family. There were, however, other disorders found in individuals in this category of well-being which, causing no continuous suffering, were consequently ignored. "Oh! I never think anything about that, it never interferes with my job". So these individuals carry their disorders about with them year in year out without attempting to secure their removal. It is hardly necessary to give examples of conditions of this nature, for they are so common:—carious teeth, running ears, chronic nasal catarrh, corns, bunions, varicose veins, constipation. Every type of disorder, in fact, is to be found in this class of individual who is unaware of or disregards his disorder. The nature of the disorders found in this category of *well-being* varies in no way from that found amongst individuals falling into the category of *dis-ease*. Thus it is not the nature of the disorder that determines whether disease will ensue or not, but the nature of the *individual's response* to that disorder.

Why then do individuals with even manifest and obvious disorder not necessarily suffer disease? The reason, well known in medical science, is that the body has an immense power of adaptation to its circumstances by which it can 'compensate' for defects and deficiencies as they arise. As long as this process of compensation is being efficiently carried out, disease and discomfort do not afflict the individual. He remains in 'well-being'. *Well-being thus represents the individual's capacity adequately to sustain a compensative existence.*

In earlier chapters of this book we have given much attention to the relationship of the functioning organism to the environment, showing how instant and versatile is the interchange that exists between them in mutual synthesis; i.e. in health. We must now turn to the organism itself and consider its own inherent flexibility and buoyancy—of which we shall later see the supreme evidence in pregnancy. The adaptability which these characteristics give is derived from the inherent endogenous reserves of living material. By means of these reserves the living machine is capable of instant self-adjustment to strains and stresses as they fall upon it. If one organ or system is smitten, other organs or systems of the body instantly contribute of their reserves thereby maintaining the equilibrium of the body as a whole. In doing so they cover up the defect, but meanwhile—and this is the important fact—the insidious consumption of the body's wealth goes on.

For many generations the use of this natural power of compensation has been the good physician's most valuable method of therapy. Is the kidney disordered? Then call in the skin, the lungs, the bowels, to sustain the burden of excretion. So sweating, purges, expectorants, etc., are prescribed as assistants in Nature's own method of adjustment. Moreover, the physician orders a 'gentle life' or a specified climate. This also is in line with Nature's process of compensation, for not only does the threatened living entity consume its own reserves in order to maintain its equilibrium, but sub-consciously and imperceptibly it *retreats from the situation* in which it finds itself, thereby shielding its incompetence from the diversity of the environment. Perhaps the great underlying truth of psychological medicine, namely that the most delicate signs and indications of disorder are to be found in psychological rather than in clinical manifestations, is only another aspect of the compensative mechanism.

In this case, as well as using up his endogenous reserves, the individual is also *using the environmental apparatus* as a compensative mechanism to preserve his entity. He circumscribes his excursion in order to maintain himself in balance—but on a lower level. Thus in compensative existence he is no longer, as in health, using all his reserves to encounter ceaseless change in the environment, but must use them to counter the changes and defects of his own mechanism. And, while the defects remain so adjusted he retains his sense of *well-being*.

As the individual moves into this compensative existence, he begins imperceptibly to weigh the balance of his reserves before taking action, and unconsciously to exclude the liability of encountering stimulus from the environment lest he be tempted to do more than he can. So, watching him in action, as in the Centre we can do so well, we see him beginning to change his ways. It is an insidious process, the individual himself usually remaining unaware of what is happening. Indeed at the height of early compensation he may even experience a sense of buoyancy—

of extreme well-being—for, having just released his gear-handle from the forward position (of health) into the quiet retreat of compensative existence, he reaps the satisfaction of 'achievement'. His desires always running parallel to his inherent capacities, he may continue unaware of his predicament even when the wealth of his reserves nearing exhaustion, he is on the brink of physiological bankruptcy.

But it might be asked, since compensation represents the body's apt use of its reserves, is not compensative existence, or well-being, akin to health itself? We must not, like the individual inured to his disorder, be deceived about this. Reserves that are occupied in continuous uni-directional adjustment of a disorder, as is the case in compensative existance, are *fixed* or mortgaged reserves. They are now no longer available for use in the ever-varying interplay of organism and environment in the spontaneity of mutual synthesis. So compensated disorder constitutes a limitation of *functional capacity* for action and hence a threat to the organism and to its parts, even though the operation of the mechanism that counters the threat is an expression of an operational capacity for health.

It is true that the more responsive and versatile the individual, i.e. the more 'healthy' he is, the more instantaneously and unconsciously will his reserves be called into play. But in proportion to the efficiency with which they operate compensating for defective functioning, they *mask* the disorder, while the individual with the disorder so cloaked continues free from *disease*. But meanwhile the disorder goes on its course silently undermining his constitution and passing unheeded from what is probably an easily removable early stage to intractable chronicity. From the point of view of the cultivation of health, our concern is not so much with the fact that this chronicity is making the ultimate eradication of the disorder more difficult—which is indisputable—as that this protracted period of limitation of function is robbing the individual, probably for ever, of his potentiality for continued growth and development; i.e. for health.

Nevertheless from the standpoint of the clinician, is not compensative existence a highly desirable condition? In the old days when very little was known of the aetiology of disease, and while there was consequently little possibility of rational therapy, to engage the body's power of compensation was the best the physician could do for his patient. Now however that the progress of medical science has thrown such a flood of light on the nature, on the diagnosis and on the treatment of disorders, any inherent mechanism tending to cloak the earliest signs takes on a very different aspect. The more slow and 'silent' the progress of the disorder, the deeper it bites into the constitution before it is discovered, making rational medical therapeutic adjustment at some subsequent period only the more difficult. As a therapy, therefore, the power of compensation must now be used with much discretion, and *only* in those cases where the disorder is ineradicable. In other circumstances compensation—stealthy thief of the body's riches—is as much a snare to the clinician as it is a hindrance to the health practitioner.

But, during the fifteen years that we have been studying health, it has been our experience that the medical practitioner and the specialist regard this state of well-being to which they return their patients after illness, as that of *health*. "Come back if you notice anything wrong" is a very usual phrase of a *careful* doctor. That is to say he is satisfied as long as his therapeutic efforts convert a decompensated disorder into a compensated one. The patient himself connives at this, for how often will he insist upon leaving his bed and dismissing his doctor the moment his compensation is re-established and *his subjective sense of well-being* re-asserts itself. While this position is in keeping with traditional medicine, it is not in keeping with the advances of modern medical science.

As we have already stated, on discharge from medical treatment all members of the Centre come up for health overhaul. Almost without exception, it is found that though the disorders for which they sought treatment have been removed, the condition of their *health* has much deteriorated. This fact is either unrecognised or ignored by the clinician ; and the more serious the disorder that has received treatment, usually the less the attention given to the re-establishment of the sometime patient. We have found that patients returned from medical care of major disorders such for instance as pneumonia or jaundice, from operations, and above all those recovering from infectious fevers, are rarely in a condition to carry on their everyday occupations to which they have been returned by the medical profession. On re-overhaul of such individuals, known to us to have been in adjustment before the onset of their illness, we have consistently found not only a general state of lowered tone, but a host of minor disorders and deficiencies such as anaemia—that is to say iron deficiency well below the accepted clinical normals—, leucopenia, avitaminosis, lowered blood sugar, muscular atony, chronic otorrhoea, constipation, bladder troubles, etc.[1] This return of the individual to work in a condition of diminished health (devitalisation) after a period of sickness and inactivity, is unscientific, diseconomic and, in the present state of our knowledge an unjustifiable procedure.

The discovery that efficient compensation is accompanied by a sense of well-being—which is not health—is, we believe, a major contribution to medical science. The features of this state of well-being, or compensative existence, that make it a menace, may be summed up as follows :—

(1) The masking of the onset and presence of disorder :—a serious concern to the clinician (Sickness).

(2) The loss of the reserves for use in mutual synthesis with the environment :—a serious concern to the biologist (Health).

It is significant that this category of persons in well-being should include by far the largest section of those examined. From this finding we must infer that these people form the unseen source from which the sick endlessly flow to occupy the attention of the medical profession and in the later stages of their chronicity to fill the ever-increasing number of hospital beds.

It thus becomes of arresting importance at this time when future developments in medical administration are under consideration, that the Centre with its health overhaul of the family reaching out beyond disease into the realm of hidden disorders, provides circumstances which bring *all* diagnosable disorders to the medical scientist at the earliest moment *he* can diagnose them, whether they are causing the individual disease or not. Our experience has taught us that ultimately *nothing short of periodic health overhaul on a national scale can lead to the rational application of medical science for the elimination of Sickness.*

[1] See *Biologists in Search of Material*, p. 69.

Diagnosis of the Early Case.

There are certain important features of periodic health overhaul as carried out in the Centre which are essential to its success as a net with which to catch disorder,—important, that is to say, from the point of view of the clinician and the Sickness Services as distinct from the demands of Health. The first of these is the approach through *the family* rather than through the individual as is elsewhere the universal custom.

Often an individual's disorder cannot be diagnosed efficiently without knowledge of his family. One symptom may occur in one member of a family and may appear trivial until seen against its family background, while its aetiology may remain obscure, even to the doctor, without a knowledge of the other individuals of that family found to have other and often cryptic signs of the same disorder. Only when all are assembled does a recognisable clinical syndrome emerge. Nutritional deficiencies are notably in this category ; endocrine disturbances afford another striking example. Personal knowledge of the constitution of the parents —that great asset of the family practitioner of the past—in these days of scientific medicine needs to be, not relegated to the scrapheap as outworn, but reinforced by knowledge derived from methods of precision capable of substantiating primary intuitive clinical wisdom. What is true for diagnosis is also in a large measure true for adequate treatment ; contact with the whole family is an essential.

An example might well be given here to illustrate the importance of a knowledge of the whole family. A family consisting of mother, father and two children and the grown-up son of the mother by a first marriage, joined the Centre. Their membership came about through the small son of 10 years who first came as a guest of another member-family. This boy impelled his mother to come and join the Centre. She came, but in great trouble knowing that she would not be able to persuade her husband, with whom recently she had not been on speaking terms, to come. No, it was as she feared. The doctor, whom she finally saw in the hope of overcoming the rule of family membership, was adamant. The mother and son could not join alone. It was then decided that the importunate son should bring his father alone to see the Centre and to meet the man doctor. The boy succeeded ; the father was won over on condition that he need only appear for his overhaul and family consultation, but at no other time. As in the Centre the members are free to come and go as they choose this constituted no difficulty, so the family joined.

At the laboratory examination the daughter of 15 years was found to have diabetes, and it was disclosed that she was under continuous treatment at a diabetic clinic. This girl's diabetes turned out to be the cause of the trouble between husband and wife, the father accusing the mother of having introduced diabetes into "his family" through her "inherited tendencies".

At the family consultation they heard that the boy of 10, on repeated examination, had been found to have a high blood-sugar, and that the father not only had a high blood-sugar but a low sugar tolerance and some sugar in the urine. Only the mother and her son by the first marriage were free from any sugar imbalance! As well as a complete reconciliation occurring between the parents when the facts were made known, there resulted a re-organisation of the family diet. Whereas before the consultation 7 lbs. of sugar were being consumed weekly—besides sweets not accounted for—subsequently this was reduced to 2½ lbs., and the imbalance disappeared from both father and son. The family consultation, in fact, became a solvent for both bio-chemical and social disorder in this family.

Our point here is that by examination of the whole family we gained a knowledge of the direction in which disorder was likely to occur in the event of long-continued habit, or in the event of any continued or unusual strain. The circumstances of the Centre made—not drugs—but knowledge available to this family, and so placed them in a position to forestall what was at least a dangerous habit for them. Only the girl had incurable diabetes; the father and son had merely a sugar imbalance which would have been considered of no importance apart from the setting in which it was encountered, but which, *in the family circumstances*, called for timely re-adjustment.

It is unfortunate that the *cure* of sickness is so much more spectacular than its *prevention*. We do not claim that the father would have developed diabetes in his declining years or that the boy would have become a confirmed diabetic, but clearly this is a case where knowledge of the whole family led to discretionary action of a valuable nature for the prevention of sickness. This case, as the reader will have noticed, had its socio-pathological aspect in the quarrel between the parents. This also was brought into solution through the health overhaul of the family.

The Centre not only provides contact with the whole family but, as a result of regular overhaul, provides a complete and cumulative dossier of the physical, mental and social condition of every individual of the family. In this respect the Centre organisation presents a strong contrast to the general practitioner service. The 'family' practitioner, as we have seen, can only contact the family in sickness—for the one exception to this, ante-natal and maternity work, is now to a very great extent lost to him, at any rate in urban practice. So he can have no knowledge of the individual until that individual becomes a 'patient'. The statement so frequently advanced, that general practice *as it stands* affords a suitable basis for a Health Service, is therefore inaccurate and misleading.

A second important point of administration is that the staff of a Health Centre must be people of high technical skill and experience in the detection of early disorder :—a subject far more difficult than the diagnosis of established disease where the patient complains, the symptoms have taken on a pronounced character, and the examiner can proceed from a definite locus of disorder. Skill in diagnosis of the presence of early disorder in the uncomplaining man in the street, like skill in other branches of Medicine, must come from practice—in this case, from continuous contact with *uncomplaining* individuals. At the present time, as we have shown, apart from the Centre organisation this type of material is nowhere available either to the doctor or to the student. No organisation other than one in which the presumed healthy present themselves for review can provide the material suitable for gaining experience in this coming branch of medical science.

Treatment of the Early Case.

When we come to consideration of the treatment of early disorder, the position is even more difficult. Treatment is not the work of the health practitioner, but it is essential to him that treatment should be adequately performed in order that the potentiality for health may be developed. Experience has shown us that there is at the present time the greatest difficulty in finding the necessary facilities for the investigation and treatment of the early case in the established Medical Services.

It must be remembered that when an individual is already incapacitated by manifest disorder, he is *obliged* to submit to inconveniences likely to lead to his cure. With an individual not so incapacitated it is essentially different. We have found that the overwhelming majority of those with disorder will gladly be treated if the treatment will not interfere with their work. Quite other circumstances than those at present in vogue in clinics and Hospitals are necessary to reassure the individual that his treatment will not force undue incapacity upon him, for existing provisions do not allow of the patient's convenience being taken into consideration. The need for special provision both for the diagnosis and treatment of this type of early 'case' was acutely foreseen by the late Sir Walter Fletcher, who recognised that the establishment of the Health Centre would involve the necessity for a special (clinical) Research Department for the investigation of the early case with which the existing medical and research institutions are not at present familiar and with which they are not in a position to deal. This need for extended clinical facilities of a new sort for compensated disorder has become daily more apparent as the Centre's membership has grown. Had it not been for the lively and sympathetic interest of members of certain of the teaching Hospitals, who were prepared to modify their usual routine to meet the necessities of an exceptional situation, we should have had to wait and to watch early and suspected disorders *slowly degenerate into frank disease* before being able to secure *scientific* medical attention for them.

The Biologist 'Doctor'.

Having hitherto had relatively little practice with the earliest signs of disorder, it is by no means always possible for either biologist or clinician to make an immediate diagnosis between health and sickness. For example, out of six cases *suspected* of cancer other than those cited earlier in this chapter, four turned out to be of a simple nature, while the other two cases which ultimately proved to be cancerous necessitated three visits each to the surgeon at intervals of not less than a month before the final diagnosis of malignancy could be established. Thus in its earliest stages even cancer cannot always be recognised as such by the diagnostician. But where periodic health overhaul is the routine, any condition which is the least suspicious can be watched by the Centre's doctor who—there for health rather than for sickness—can make his observations without arousing apprehension, perhaps unnecessarily, in the individual's mind. As Health doctor, he is in a very different position than would be the doctors of the Cancer Clinics which it has been suggested should be set up for the early detection of that disease.

It is necessary that the (medical) staff of a Health Centre should be not only skilled in the detection of early disorder, but also that they should be conversant with the wide range of variation in the expression of function. The biologist has to learn to diagnose *health*. Without this, any variation outside the limits of the clinician's 'normal' will, often erroneously, be taken for disorder, with the result that the individual will be rendered an invalid without justification.

An example of such a situation can be seen in a man with a raised blood pressure. At his first overhaul, at 39 years of age, his blood pressure was 145. At his second overhaul the following year, it was found on repeated examination to be 160. During the year the man had changed his habit of life, and from being a manual worker and active boxer in his leisure, he had become a sedentary worker with no time for his hobby. But with the change in his circumstances he had not changed his diet nor his food intake, so had begun to store fat, which he had previously worked off in daily turnover. This explanation of the findings at the family consultation not only put the man wise to the real position, but also his wife—who was in the habit of pressing her good dinners upon him. With one or two intermediate laboratory examinations and with the co-operation of his wife the adjustment was effected to the satisfaction of all concerned, and there was no further rise in the now adjusted blood pressure. But the older an individual grows, the more inured does he become in his tendencies and habits, and thus the less versatile in spontaneous adjustment. Another ten years of raised blood-pressure and at 50 this man would probably have acquired an irreversible hyperpiesis—by that time an established clinical disorder. But to diagnose it as a case of high blood-pressure at 40 would have been a grave error liable to induce psychological disorder with a socio-pathological sequel not only for the man himself but for the whole family.

Judgment that can only arise out of training and experience is essential in the 'biologist' doctor conducting the periodic health overhaul. It is not work for the novice. There is no reason, however, to believe that it is less easy to acquire the necessary competence in biological practice (health) than in clinical practice, if the student during some part of his training is presented with suitable material on which to begin the study of health. At some time during his training he must in future have the opportunity of learning the laws of health and of seeing and studying the healthy, as well as—and as systematically as—he learns the laws of pathology and studies the sick. It is clear that study of the healthy cannot be pursued in the Teaching *Hospitals*. Both the material for study and the training necessary are different for the two branches—Health and Sickness.

The Effect of the Removal of Early Disorder.

Certain queries likely to have arisen in the reader's mind may perhaps call for answer. The first is—Can Medicine really help the man in the street before he is driven by disease to seek the help of the doctor ?

In adults in whom disorder was removed, whether it were a grumbling appendix or an unrecognised goitre affecting the heart, breathing or nervous system, the relief was instantaneous. "Now I know I have had something the matter for a long time", is a very usual remark as the individual begins to resume the activity of his earlier years. And, *the earlier the removal the greater this sense of relief*, for it is likely that the subject of the disorder will still be resisting a retreat from his environment although he may already have been using his reserves for compensating the disorder. Thus the subjective effect on the individual offers an immediate answer in the affirmative to this question.

But, we know from the nature of 'well-being' that it is unwise to accept subjective evidence alone as an indication of progress towards health. Have we any objective evidence of improvement following removal of disorder ?

In the Centre where day by day the action of its members was visible to the observer, we had abundant evidence in *a change in the individual's action* associated with removal of his disorder, to corroborate the subjective evidence of the individual himself. For example, four days after the efficient de-infestation of a child with worms, a change in that child's actions was apparent to Staff and parents alike. It was manifest in increased power of concentration, better co-ordination of his action, in a change in his attitude to his parents and an increasing ability to associate satisfactorily with his playmates. So consistently was this observed in the case of worm infestation,[1] that assistants not medically trained grew to be able to report—at some later date—that the child should be re-examined for a recurrence of the trouble. The accuracy of their observation was confirmed by re-examination of the stool with the discovery of renewed presence of the ova. Observations of action are however difficult to record consistently, and at present defy measurement, so that they cannot as yet be stated in any statistical form. There is room for a great deal of study in this subject now that a field of observation has been created.

Again, it would seem easy enough to take families at their first periodic health overhaul and compare their condition with that found at subsequent overhauls, and note the diminution of disorder found in the latter. But such a comparison is by no means a simple one. Cancer might have been discovered in an individual at first overhaul and subsequently removed, but on returning at second overhaul with no sign of recurrence of the cancer he might be in the middle of an attack of fish poisoning. Both these conditions—one grave, the other transient and comparatively trivial—take him out of the category of those without disorder, and leave us without evidence of a change in his health. How difficult, then, as yet to give any useful *statistical* picture whether based upon subjective or objective findings.

Here is another example of the kind of dilemma encountered in attempting an assessment of this order. A family of four at first overhaul have certain disorders, e.g. mother: varicose veins, obesity, bunions and an iron deficiency (70% haemoglobin); father: chronic otorrhoea and carious teeth; girl of 12: constipation, rough skin and an iron deficiency (75% Hb); boy of 16: acne, enlarged tonsils, blepharitis and carious teeth. All have signs—different in each case—of vitamin deficiency. The family is keen to be 'fit' and takes trouble to be rid of the disorders that have been found—none of them usually called serious. They all decide to take a course of vitamins. Three weeks after

[1] For incidence of worm infestation in families examined, see Appendix VI.

beginning the course they feel much better. Six weeks later the father develops a boil. Three weeks later still the boy develops a series of boils. They are miserable with the painful condition. Are they now worse or better ? How can the boils be assessed against the constipation, acne and blepharitis that have disappeared ? One is acutely painful ; the others they took no notice of. If they continue with the vitamins the boils subside and no further crop appears. They again feel well, better than ever. But had they stopped the vitamins when the boils came, gradually the original painless symptoms of avitamonis would have reappeared. The explanation for this seems to be that the vitamins raise the resistance of the body to germs of low virulence which in his initially low state the individual had been tolerating. This toleration is yet another aspect of compensation. Once the resistance is raised, the body will no longer tolerate the invading germs and a fight ensues—emerging as a crop of boils. When raised still further, the invaders are permanently cast out and no further boils ensue. Tolerance in this case is a manifestation of failure to function ; intolerance a sign of growing health—of functional action.

But supposing for instance that the family had delayed taking action for nine months, then at the second overhaul they would have been in the middle of their attack of boils. It would be equivocal to record the family in its moment of pain and discomfort as in better health than at the first overhaul, and yet inaccurate to record them as in less good health ! The ground on which we tread here is treacherous and shifting. It must remain so until the study of Function has disclosed methods of measurement of health distinct from the clinical methods of assessment of disorder. It is with *methods for the measurement of health* that the Peckham Experiment is specifically concerned.

Then again many disorders are removable with great advantage to the individual concerned, but the underlying diathesis that caused the disorder may remain—ineradicable. Endocrine disorders, some allergic states, rickets, are of this order. This means that to achieve health nothing short of attention to the breeding of the next generation is of any good.

The investigation of the effects of removal of disorder in its early stages is one which essentially needs the elapse of years for its completion. Up till now we have had too short a time to come to any satisfactory means of measurement of health. But we have had more than enough experience to indicate that routine periodic health overhaul in conjunction with an environment capable of the subsequent educement of function such as the Centre provides, is a practical measure of immense value in moving towards health. We must already claim that enhanced vitality, and a wider measure of freedom from sickness, can be attained through periodic health overhaul of families in the circumstances provided by the Centre.

Whether such results would be obtainable from periodic *medical* overhaul carried out merely with the intention of disclosing disorder and accompanied only by clinical measures for removal of the disorders *in a situation where there was no possibility of changing the nature of the environment out of which the disorders arose*, is another question. It is one we are not in a position to answer, because membership of the Centre which led to overhaul, also inevitably implied contact with an environment capable of educing function after cure, so that the liability of 'seven devils' entering in was in fact minimised. Knowing the inertia of habit, we suspect that this is a very important proviso and that if advantage is not taken of the moment of release to present the individual with circumstances which allow of a more fully functioning existence, relapse will easily follow the initial rebound towards health after removal of the disorder. But that would be only another argument for the establishment of *cultural* Health Centres as the setting in which to carry out the periodic health overhaul of families—which we have already shown to be essential for the rational conduct of modern scientific medicine.

Costs.

A second question which may arise in the mind of the reader is :—But even though periodic health overhaul as carried out in the Centre would lead to removal of disorder earlier than is now possible, would not the cost be so great as to make it impracticable ? A carefully checked estimate of costs based on our first three years' experience shows that the cost of overhaul need be no more than 1s. 2d. per week per family.[1] In consideration of the benefits to be gained from such a procedure, this sum

[1] See Appendix V. *The cost of the periodic health overhaul.*

compares very favourably with, for instance, the cost of the School Medical examinations which are far more cursory, less frequent and include no laboratory investigation in the routine examination, providing only for certain tests to be carried out *after* the suspicion of disorder has been aroused by clinical inspection of the child.

The emergence of the Centre as a type organisation would seem thus to be of arresting importance at this juncture when re-organisation of the Medical Services of the country is under discussion. The Peckham Experiment has demonstrated that periodic health overhaul is an efficient means of overcoming the present deadlock. It effects easy, natural and continuous contact between the unsuspecting man-in-the-street and the skilled diagnostician thereby bringing to the man-in-the-street the fruits of modern advances in medical science, and to the medical profession the optimum conditions for the exercise of its ever increasing skill. Nothing short of periodic health overhaul of the family on a wide scale can meet these two primary needs for reduction of the national sickness.

3 HEALTH

It will have become obvious to the reader that crucial though the above conclusions are for the rational care of the sick, they are merely incidental to our main purpose which is the study and promotion of Health. To that subject let us now return.

Just as the prism disperses sunlight into a spectrum of primary qualities or colours, and of primary quantities or wave-lengths, so the Centre technique *as a whole*—not merely the periodic health overhaul—is a prism that separates out for us three categories of existence: Health, Well-being and Disease—or Living, Survival and Dying.

It is health we set out to study, and it is useless to attempt to apply measurements for health, i.e. functional standards, to those who are already disordered or diseased. The biologist must concentrate attention upon those in whom no disease or disorder can be found. He must cultivate the leaven in the lump; that is to say the Living in society.

Now it is common to assume that those who are not sick are healthy. Not only is that far from being the case as we have shown, but such complacency with clinical or sickroom standards of fitness is dangerously misleading. This fact, which we have been pressing on the public for fifteen years, is now becoming widely recognised. Substantiation comes from other quarters than our own. Nutritional surveys, for example, have demonstrated that clinical standards do not constitute any criterion of health. In the school medical service where the doctor is not reviewing 'patients' but uncomplaining individuals, the school medical officer often finds that his experience is at variance with accepted clinical standards.

But what standards other than the clinical ones are available for the assessment of health ? There are two. One consists of the very carefully established standards adopted by the physiologist. These are directed to the assessment of the efficiency of the organs of the body and of the efficiency of the body as a whole *for a given purpose.* Those physiological standards correspond roughly to the bench-test standards of the engineer.

The other type consists of functional standards. These are as yet very tentative where indeed they exist at all. This however is to be expected, for the study of function has only just begun. As we saw earlier in the discussion on function in chapter i, functional standards can have no analogy with physiological standards—the bench-test type—and not even with those analogous to the 'road-test' standards of the engineer—the physiologist's tests for determining the working efficiency of the assembled machine, for all such standards are directed to assessing an *objective efficiency.*

In Biology (function) the standards can only be based upon a *subjective efficiency* : whether, for example, the tree is an efficient expression or manifest of the seed from which it springs ; the man an efficient expression or manifest of the child he once was ; whether society is an efficient expression or manifest of the families from which it springs. And so on.

What, then, are the facts as determined by the use of these two sets of standards and methods ?

Working on a physiological basis, in our survey of 1,206 families we find 9.2% of the people without defects or deficiencies of a tangible and measurable order—and this figure includes infants who have not had time to acquire the chronic disorders derived from ingrained habituation.[1]

[1] There were 358 individuals found to be without any discoverable disorder in a total of 3911 examined. See chart I (*frontispiece*).

But this does not mean that this 9.2% are actively fit and healthy, i.e. that they fall into the category of the *functionally* efficient. For example, a youth may have a normally working heart, physiologically efficient, but when called upon to perform a simple but unusual task, it flutters and staggers under the stimulus. A good heart—but not co-ordinated to meet the emergencies of everyday life without being upset in its rhythm or disturbing its economy of effort! Or again, a good heart and good lungs and good musculature may have been trained to some special task—running or gymnastics, for example—and may yet fail to meet some simple emergency with becoming calm co-ordination. An A1 Metropolitan policeman faints as his wife is carried off to her confinement, for example. A man habitually trained to muscular effort falls down dead suddenly on slight exertion in the prime of life. 'Athlete's heart' affords an extreme example of this order. His response to effort in the physiological laboratory the week before may have been of a high order of physiological efficiency. A week later his efficiency for *living* is demonstrated to be nil.

Physiological normality in fact does not necessarily go hand in hand with functional efficiency. This has been recognised for years now in the assessment of fitness for flight in air pilots, whose functional co-ordination should be of a very high order. It is just this functional co-ordination of action that marks the distinction between *intrepidity* and *foolhardiness*. Indeed, foolhardiness, which usually arises from a consciousness of physiological efficiency, may cover a multitude of functional inefficiencies. Foolhardiness, often praiseworthy in itself, slaughtered the flower of the physiologically fit in the last war. It was this fact that inspired Flack to inaugurate the Air Force Research into the assessment of fitness on another basis—a functional basis. Alas! Flack's enthusiasm died with him, or the science of Biology might have been further advanced to assist in the present crisis in national affairs, to meet which intrepidity is the prime need. The functional characteristic of the intrepid is their power spontaneously to deal with events however unfamiliar, however foreign—for they are responsive to any and every change, and not merely to foreseen and pre-judged changes.

Intrepidity, it would seem, now but rarely springs spontaneously from the social soil. We have reason to suspect that this is because there is *little life* in that soil. Hence the despairing search for leaders in every walk of life. When we say a man is a natural leader, we are expressing the fact that he has an *inclusive* (subjective) co-ordination of his reflexes and sensations, i.e. that he is both physiologically and functionally healthy. It is often said that the leader is born, not made. This assumption is in fact a confession of our ignorance, first of the necessity of cultivating, and second of how to cultivate this inclusiveness ; it is also a confession that the methods already tried—the methods of training—have failed to produce leadership. While the assumption is wrong there are, however, good grounds for it, for, as biology is rapidly teaching us, the foundations of health are laid down in foetal life and infancy and built up through specific nurture during childhood and adolescence. That is to say, health is not merely a genetic characteristic ; it is also a *nurtural characteristic* and, as such, is inevitably handed on from parent to child. So, a healthy family is more likely than an unhealthy one to create an intimate environment favourable to the acquisition of health, and in that sense the odds are in favour of an inherited '*tradition*' of health. But that is not what is usually implied in "born not made".

Health and intrepidity are, then, not merely inherited. From the beginning of life (i.e., conception) onwards, by the assimilation of vital nutritive factors from the parental soil—the physical and social environment—they are acquired through the nurture that proceeds from the functional organisation of the family.

So for the attainment of health, the importance of making use of biological standards which take account of the organism and its environment and of the relationship between the two is obvious.

Applying such biological standards as we at present can, and from such observations as we have been able to make on the 9% of our populace who are physiologically normal, it appears that but few even of these can be classified as functionally fit and healthy. Though physiologically normal, they are in a state of suspended function, or 'survival'. They have the capabilities for action but lack the co-ordination to make action effective. Of this, they are often all too conscious ; we see them shyly sinking into inactivity later to become apathy, or, urged on by the consciousness of physiological fitness, adopting a neck or nothing foolhardiness and rashly attempting the barely possible.

How is it that this can occur? What factors are present or absent from the environment of these people that inhibit growth and emergence of capacity and their enjoyment of their state of physiological fitness? We have already at the Centre succeeded in demonstrating that it is not due to any lack of potentiality. Indeed, the machine may be super-charged with physiological virility. The cause or causes lie elsewhere. Are there any clues that we can follow?

We shall record later that a considerable proportion of young individuals—children and adolescents—who on joining the Centre are functionally inefficient, begin to blossom into co-ordinate action, throw off their apathy (often mis-called 'stupidity') and dissolve their shyness in geniality, or translate rude impetuosity (often mis-called 'viciousness') into calm deliberation, *once their families begin to be integrated into the social life around them* in the Centre.

The importance of this observation is that it has appeared to be the movement of the *whole family* into a social sphere of action that has given the initial momentum to the latent capacity of the individual child. In other words, it is *familiar nurture* that they need, for of 'nature' they have plenty.

In confirmation of this, we also find that those few individuals who do exercise their capacity in co-ordination, the intrepid, are members of families with a marked social integration. And further, that among the functionally inefficient, those whose families cannot fit themselves into the social life tend to show no such improvement.

We have in earlier reports drawn attention to the fact that the effectiveness of the existing methods of education has been observed by us to vary directly with the degree of social integration of the family from which the child comes. In watching the course of children leaving school at fourteen, we find that their progress tends to be spontaneously continuous, even without any continuation school assistance, in families with a high social integration; whereas it ceases almost abruptly, even with every institutional assistance, in families with little or no social integration. By seventeen, children of this latter group may have almost forgotten how to read and to count. It is the *functional efficiency of the family* which produces the 'digestive ferment' and is the formative and synthesising factor in education; and not the mere knowledge purveyed by training. Like the process of nutrition, Education appears in fact to depend on the *efficiency of utilisation* as well as, and as much as, upon the nature of the diet available.

It is not of course claimed that the integration of the family, within itself and with society, is the only means of activating potential capacity. The absence of integration may, as we know, to some extent be remedied—though it cannot be *cured*—by training. We claim however that integration of the family developing in mutual synthesis with its environment will prove to be the biologically economic way of developing human potentiality—the way of *health*. If what we as biologists regard as the grown mature organism, the family, can find the wherewithal to operate on and to synthesise, then the immature individual growing up within that family will participate in that experience and in so doing will *spontaneously* acquire a natural appetite for, and power to, digest his own experience as that unfolds. The family, in fact, is the natural 'organ' for predigestion of all experience which reaches the child from the environment, not only during gestation and lactation, but until the young leave the nest.

So, functional efficiency is something which is derived *from within* the plant or the animal, in circumstances where the soil in which it is growing contains the essential nutritive elements; i.e. where the soil is as living as the growing plant or animal itself. Everywhere the biologist finds Nature using specific nurture, i.e. familial nurture, to lead the young on to the ability to digest and utilise the non-specific and the foreign experience. It is *the family* then that Nature places at our disposal, not only for the propagation of variations in the species, but also as *her instrument for the cultivation of functional efficiency*.

So health does not demand education of the individual, nor education of the populace—the two accepted and popular methods—but education *of the family as a live functioning organism*.

Health in the past has emerged sporadically, where good seed has fallen into good soil. The time for that is now past. The seed may no longer be cast to the wind to settle where it can.

Tilling the familial and social soil of man is becoming a science and art to be acquired with all the assiduity—and more—given by man to the study of physical phenomena and to the study and cultivation of his plants and beasts. Whether it is sought in virgin ground or in the weeded fallow from which disorder has been cleared, Health is a cultivator's problem, and that cultivator can ultimately be no other than the biologist.

CHAPTER VII

NEW MEMBER-FAMILIES

A FAMILY has joined and they have had their first health overhaul. How do they become involved in the social life of the Centre and begin to take part in its activities ?

Coming and going for their various appointments for examination, they have passed through the cafeteria, and seen what is afoot at different times during afternoons and evenings. They have probably had tea there, or the parents may have had a cup of coffee or a glass of beer after the father's overhaul. They know the member who first took them round, and if she is there she recognises them when they come in. She is pleased that they have joined and introduces her husband to them. They have probably also spoken to one or two people in the reception room ; asked the way of another member ; and of course they know the biologist on duty about the building whom we have already called the 'curator of the social instruments' [1] She with the 'doctors' met them at their enrolment, and they know that she will give them any information they want about the various activities they see going on around them.

If there are children in the family there is no hesitation. The children want to do something at once and have probably waited impatiently for the moment when they may begin to learn to swim—for they may not go into the bath until their overhaul is complete. Then proudly they bring the curator a ticket from the doctor, and find a time when, often with other learners, they can have their first lesson. When that day comes the mother goes to the learners' bath to watch her boy and so falls into conversation with another mother as they both watch their offspring struggle in the water. Or, in a family where there is a baby, the mother will have an appointment to come again in a week's time for the infant's weekly visit—and a baby is always an introduction to other members.

The husband has perhaps met a man he knows at work, and together they go off to have a game of billiards. So through action he too learns his way about and tells his wife how admis-

[1] See Chap. IV. p. 78.

sion is gained to the various activities by the grown-ups. The 'key' described in chap. IV not yet having been installed, up till now there have been various ways of making payments for these activities; either by ticket bought at the cafeteria cash desk (e.g. for swimming and entry into the gymnasium); by weekly payments collected by secretaries of various intramural clubs (e.g. badminton, billiards, table tennis, fencing, discussion group, etc.); by collection by authorised members of groups or committees such as the Concert Party, the organisers of the Christmas Party, and so on.

Each approach by the new member involves contact with some individual with whom a mutual transaction inevitably takes place. As, one by one, such contacts are made, acquaintanceship grows from a basis of common action and interest. These first actions are like the tender new root hairs put out by a plant in new soil, presently to sustain new growth in the whole organism. Beginning with threads slender as are these first tentative though lively contacts, the family unconsciously begins at once to weave its own contribution into the gathering pattern of social integration of the Centre.

Is there no one then to act as Centre hostess: no one to make introductions and help new members over their first stile? No—only the sight of activity all around, which in itself constitutes a continuous invitation to the new family. As biologists anticipating and planning for health and virility in the members of the Centre, we are looking for evidence of *spontaneous* action in new circumstances; we must, therefore, hold our hands and be patient. By ourselves stirring action in the members we should fail to find and to see what we are looking for.

It is true that by pursuit of this method we have lost some families as members. It is also true that on those occasions when we have taken the reverse course and used the method in vogue in most social clubs of deliberately introducing new families and individuals to established members, no better result has followed. The shy are only too apt to be confirmed in their shyness when sitting down to talk to strangers with whom they have no link of any common pursuit or interest. By that method we have found that the dependent family is only confirmed in its dependence at the very moment of having taken the bold step—perhaps its first social venture—of joining the Centre at all! Were the family to be offered a social crutch on crossing the threshold of the Centre we should be robbing it of the very chance it has made for itself of taking the next step in the mutual embrace of a new environment; that is towards its own health.

The Incentive to Action

The reader will recall that the task we set the architect was to provide a building so planned that the *sight* of action would be the incentive to action. Four years' experience in the Centre has established the postulate of the potency of vision and propinquity as an effective invitation to action for people of all ages. But it must be remembered that it is not the action of the skilled alone that is to be seen in the Centre, but *every degree* of proficiency in all that is going on. This point is crucial to an understanding of how vision can work as a stimulus engendering action in the company gathering there. In ordinary life the spectator of any activity is apt to be presented *only* with the exhibition of the specialist; and this trend has been gathering impetus year by year with alarming progression. Audiences swell in their thousands to watch the expert game, but as the 'stars' grow in brilliance, the conviction of an ineptitude that makes trying not worth while, increasingly confirms the inactivity of the crowd. It is not then all forms of action that invite the attempt to action: it is the sight of action that is *within the possible scope of the spectator* that affords a temptation eventually irresistible to him. Short though the time of our experiment has been, this fact has been amply substantiated, as the growth of activities in the Centre demonstrates.[1]

The reader will now appreciate that it is no accident in the design of the building that to reach the reception vestibule and consulting rooms for the initial enrolment, it is necessary to walk through the cafeteria with full view of the swimming bath and other activities.[2] In so doing the enquiring family all unconsciously taste the full flavour of the buoyant life they are moving towards. Once joined, they are surrounded by many activities of which they may never have felt the attraction before—a very different situation from that of the man who joins a billiards club

[1] See Appendix I.

[2] Photographs 3 and 4, p. 53.

or a dramatic group urged by an already established interest in billiards or in acting.[1]

In the Centre the design of the building makes it very difficult for any but the most inert to sit day after day at the cafeteria tables overlooking the swimming bath and not eventually succumb to the insidious urge themselves to join in the activities. We have aimed at making entry into every activity as easy as possible, not only for those already skilled, but for the shy beginner feeling the first dawning of interest, and who is so easily discouraged by the expert and the professional. That this should be so is the 'curator's' special concern. He or she must see that entry into the bath, for instance, and friendly instruction in swimming for the older members and others who need that assistance, is easily available and as unintimidating as possible.

The place of Intramural Clubs

The initiation of clubs or groups for the pursuit of any adult activity is in the hands of the members themselves. The curator is responsible for the allocation of floor space as between the different activities and is *ex-officio* member of all committees, Her constant concern is that nothing should impede the emergence of fresh interest and that all activities should as far as possible be continuously available to the unskilled as to the skilled; to the shy as to the bold.

We have found that any intra-mural club such as a billiards club, valuable as it is as a focus of interest and activity within the Centre, has an inevitable tendency towards exclusiveness, not necessarily because of its members' wish but in virtue of their good play. The shy and indifferent player cannot bring himself to play with these experts, and the club—a closed coterie of members becoming increasingly proficient—tends to draw less and less on new members of the Centre. This can be overcome by keeping the majority of the tables open for casual use by all (on payment of the appropriate sum per game), reserving certain tables only for the use of the billiard club (whose members pay

[1] In this situation, with a membership composed of families led to join for a wide range of reasons, it has been disclosed to us how few among our members had any leisure interests sufficiently strongly established as to have led them to join any specialised club or organisation—whether based on sport, politics, or anything else. For a discussion of this point, see *Biologists in Search of Material*, p. 37 et seq.

a weekly subscription collected by one of themselves). The club itself benefits by this arrangement for now it can be continuously fed from those among the casual players who show interest and capacity. The fact, too, of there being tables reserved for the skilled is an attraction to those already skilled to join the Centre. This works for activities of all types as each in turn becomes established.[1]

Those of our readers who have followed this experiment from its first beginnings and are familiar with *The Case for Action* will recollect that in that short forecast we made certain postulates with regard to group organisation, procedure and routine in the larger Centre. We have now abandoned all deliberate methods of organisation in favour of a more individual, free and spontaneous development. This change of method resulted from experience gained in the first few months during which time we discovered that children have a great volitional wisdom if allowed to exercise it in a *social* setting among their elders. In circumstances where they are not starved of action, it is only necessary to place before them the chance or possibility for doing things in an orderly manner for them to grasp it; they do not need, indeed they resent being either herded, coaxed or guided into action. And in the circumstances of the Centre neither are the adolescents—usually considered a 'problem'—nor the adults any less capable of directing their own action. Thus, the attempted promotion of any form of stereotyped organisation based on 'leadership' was early discarded and in the subsequent chapters of this book the reader will find a notable absence of deference to the modern clamour for *leadership*. Our members have already taught us that leaders require no training; they emerge naturally *given the right circumstances*. In the Centre the visitor is generally very surprised to learn that what he sees before him is spontaneous action and not the result of programme, persuasion or regulation. It is the conjunction of order with spontaneity that faces him which so repeatedly draws from him a question about leadership.

[1] As evidence of the social value of groups of the skilled formed into a more permanent 'club', we can cite the custom that has arisen among the various clubs of holding 'Club Socials' within the Centre. These generally take the form of supper parties of some 100 to 200 people often followed by a dance, cabaret, etc. On these occasions the members make themselves responsible for the preparation, catering and clearing up after the party, in accordance with the *self-service* principle of the Centre.

Let us follow rather more closely the history of the initiation in the Centre of these activities, seemingly so casual. Almost without exception any group-action, such for instance as the drama or dancing, has in the first place been initiated by some rather unbalanced, uncritical enthusiast, who appealing to the management to meet his desires, was thrust back upon himself to collect, if he could, other persons infected with a similar enthusiasm. In many cases the initial enthusiast was so precipitate that he failed to gather adherents; or he might, by dint of perhaps some weeks' or months' propaganda and search, succeed. This little group, then, operating in the open forum, either failed or succeeded in creating a more or less general, though perhaps hesitant, interest on the part of other members who came into contact with it in their casual round of the building. And so, slowly but surely, some demand for this activity was built up from among those who were caught by and shared in the primary enthusiasm. Too often at this point of success, the curator was approached by the original enthusiast with a request to create a permanent official committee. The curator, however, having watched the growth of this activity, had noted that the great body of adherents manifested no serious desire for any fixed organisation—and indeed showed remarkably little gratitude to the enthusiastic initiator of the group. The outstanding fact was that most members of the group had merely been learning the first simple steps in a process, whereas the enthusiast had all the while been attempting to polish his relative perfection. Sooner or later then, this distinction between the ordinary members and the initiator led to disappointment in the enthusiast, who generally went off with perhaps two or three relatively proficient exponents to the neglect of the ordinary members. With this arrangement both sections seemed to be well satisfied. The unskilled did not now fall away because of the enthusiast's neglect of them but proceeded to throw up from among themselves a new leader more in conformity with their own lesser capacity and skill, while the original enthusiast, unimpeded, was in a position to further what ability he had with his new found companions. Much experience of this sort has taught us that all committees should be *ad hoc*; for self-appointed leaders come and go. Hence to stereotype an organisation on the basis of any primary enthusiasm is usually to sterilise the spontaneity and effort of the *ordinary* person, and often to thrust him back into inactivity. The permanence of a committee in fact creates a vested interest and a specialism, and curiously enough leads to the creation of what might be called a 'class consciousness'--in this case a 'club consciousness'—and that tends to *ex*clusiveness. So the rule in the Centre has come to be that no committee shall be the ruler and director of experience nor be allowed to create a vested interest.

The adoption of these principles of *ad hoc* initiation and control of activities, much as it has annoyed many of the more conventional and habituated members with fixed ideas of how things should be done, keeps everything fluid, malleable and inclusive. And strangely enough, the result is order and not chaos. The important point is that it raises the general level of the community, instead of merely fostering the achievement of a high standard by a few specialists, while it does not prohibit the gifted from creating the circumstances that will give them proficiency. Far from the specialists being discouraged, we found that they were led to join in protagonism with their peers rather than to exercise their vanity on those less skilled. Moreover, the very presence of the building with the facilities it holds out and the freedom of action it promises, tends to draw those to join as members who already have some skill.

It will be noted that this rule of *ad hoc* initiation and control of activities is an extension of the principle of individual freedom which we have already seen to be inherent in self-service. It is a common experience in any committee for the complaint to be voiced by its members that their service to their fellows interferes with their own individual action and is unrequited. The inevitable corollary, that the club members' action is limited and that they are ordered about by those in authority, is all too well known. For a more scientific consideration of these problems we would refer the reader to a discussion of 'order in anarchy' in *Biologists in Search of Material*.[1]

The Centre as an organization is 'alive'

The building, as we have said, was primarily designed to be furnished not *for* its members, but *by* them and their actions within it. There can be no doubt that, together with the richness

[1] p. 40-42.

and diversity of Nature—who for her part repeats nothing and changes from day to day, from season to season, from place to place, in weather, soil and oecological sequence—man's own ingenuity and inventiveness is a contribution of the highest importance to the diversity of his environment. The products of his necessity and exuberence, his goods and chattels, his playthings and his tools, his science and inventions, his poetry and music—all complete this diversity and in turn through mutuality of action become the food for his own further development.

The Centre then, with its ever changing, varied and extending chances for each family that joins, is a building designed in every aspect to afford a stage upon which the *action pattern* of family function can proceed to develop spontaneously at all times. In essence, with its rough framework of perhaps even crude chances for every member-family, it is not unlike the wax sheet with which the hive-keeper presents the bees. According to their strength and according to the food available to the hive, so the insects pull out the wax into 'cells', specific in architecture to their breed, there to deposit their honey. But it is the bees that make the honey in their own time; not the hive-keeper with his wax frame. And so with the Centre—it merely gives the frame; the family secretes the specific honey, each family-organism contributing to society its own peculiar flavour and quality. In laying down its own action-pattern it creates a novelty of its own which adds to the richness of the society of which it is a part, and from which in turn it draws further sustenance—'pabulum' for its future specific development.

Perhaps, for those readers who have never visited the Centre nor felt its natural spontaneous atmosphere, a picture of how it can become a field of opportunity for a young family, in which situations such as we describe occur easily and naturally, would be useful at this point.

A vignette of life in the Centre

Here is a young couple, four months married, who have just become members. They are shy, diffident, have no friends in the district—only mother-in-law. They know no one in the Centre. They come to their overhaul and then one Saturday afternoon to their family consultation. After this they go downstairs together and have tea in the cafeteria. They do not hurry for there is much to see ; they watch the swimming bath. There are other young couples enjoying a swim together, perhaps one father has a baby of two or three years on his back. The husband of our young family can swim. He wants to go in. He wishes his wife swam too. He registers a determination to come next Saturday with his swimming suit and have a dip. They cannot linger any longer over tea, for they feel they are becoming conspicuous—sitting doing nothing. They get up, and loath to go, saunter round the bath. At the far end they stop again ; looking down through the window into the gymnasium they see 25—30 boys and girls of all ages swinging on ropes, climbing, jumping.[1] "Why Fred ! it's just like the monkey cage we saw at the Zoo", she says. In one corner on the mats are two young men—"Coo, they're good", says Fred, for they are fine athletic fellows practising somersaults,—"That's the thing one chap was doing from the diving board in the bath". These grown-up athletes do not seem disturbed by the children who for the most part are absorbed in their own concerns, though a few watch admiringly from their point of vantage on the ladders or cross-legged on the floor. As the young men go on to even more difficult feats the children sit quiet as mice and their eyes shine. The men pause for a rest and turn to the small audience. "Do you want to do it, Jimmy ?" —and so the two give the little group their first lesson in aerial somersaults.

Our young family watch for another quarter of an hour, and then their attention is attracted by a strange noise of drums, and a primitive music rises behind them. It is the children's band in a corner of the Long Room. All ages, 3 to 13, are assembled ; each takes some string or percussion instrument or for the littlest there will be a wooden cylinder full of shot to shake to the time of the tune. There is a pedal instrument to tell the company by a colour indicator what chord to play. For the rest the rhythm leads them. They put up a reasonably good show ; someone begins to sing and in the end everyone has joined in. "It's really rather fun". So our young family passes away down the corridor and finally leaves the building.

Next Saturday, a little later, about 7 o'clock, they come again. The girl is very neat and smart this evening. The man has his swimming suit. He has bought a ticket and is going into the bath.

[1] See photographs 24 and 25, p. 61.

The girl, shy at being left on her own, goes to the quietest side of the corridor overlooking the bath and watches eagerly for her man to appear in the bath chamber. He comes up; she shifts in her seat, smiles and waves to him. He too is shy—almost too self conscious to return her greeting; he slips unobtrusively into the bath from the side and swims. Except for a few learners everyone dives in, but there are lots of people there, and he is not confident of his style—better just swim about.

While she looks on, the girl sees two couples—older people than they are—with their children come up into the bath. Mother is very stout; how can she have the courage! But somehow she seems unaware of it and bears herself with remarkable assurance. Wait, she is actually going to try a dive—"Well!" She does, too .... "Perhaps I could learn to swim? No, I should never have the courage to go in that bath with everyone looking".

Here is her husband back, feeling very bucked and refreshed. "You'll have to come in, Flo". But Flo quickly—"Let's have something to eat", and so they go to the cafeteria. They help themselves and sit down at a table overlooking the bath. People are beginning to come in rather fast. Our young couple are lost in the growing crowd—"It's all right ...."

The two families the girl had noticed in the bath have come out. Here they are and in two minutes they have put two tables together in the cafeteria and all sit down to a high tea. They are very friendly; all know each other well. They are talking about what they will do to-morrow. One of them has a small car. They are going to the country to some place they call "the Camp". It must be something to do with the Centre.[1] Anyhow they are all going together. And now they have finished and are getting up to go home and put the children to bed.

"I overheard from a fellow in the changing room that there's a water polo match on to-night. Let's sit here and wait for a bit". At 8.30 they still have courage to sit where they are, our shy young family. Then someone comes and draws nets along the inside of the bath window. Round the bath chamber the crowd is getting denser, every table is taken and people are standing three deep behind. You can't see the cafeteria counter for people. Here they come, a visiting team—red caps— against the Centre—black caps. Tremendous excitement in which our young

[1]See note on Camp—Appendix II.

family join .... and they find themselves talking to the people next to them.

An hour later they have moved round to the other side of the building. The Centre band is playing—"there's the young doctor playing the double bass"—and in the midst of the dancing they take courage. They too dance—just together—for a good hour.

There is an informality about the life of the Centre which permits of the spontaneous emergence of any situation at the tempo natural to each family. Those unused to move in any society find themselves gently lapped by the tide of its action.

A very high proportion of the families we have encountered are characterised by the absence of social contacts of any sort save those of the relatives of husband or wife—and these last are generally too close and too insistent to operate as anything but a restricting factor. One of the outstanding features of the populace from which our members are drawn is their inability through this lack of social excursion to take advantage of any new situation that presents itself. This, membership of the Centre changes. To its member-families it becomes an open door leading to fuller, more varied and unknown aspects of life—a door beyond which there lies adventure. Each family on joining is free to pass straight through into new territory, or equally free to stand against the doorpost and merely watch, not daring to think that inevitably adventure will draw it on—in its own time and according to its inherent capacity to go forward.

CHAPTER VIII

THE FAMILY GROWS

1 CONCEPTION

Six to eight months have passed. The picture of the young family we followed in the last chapter is now very different. The girl, in a smart little pair of shorts and red shoes, comes briskly in by herself at 5 o'clock. She is going to take a tap dancing class for some children. Her husband is now in the water polo team. They both belong to the Concert Party. They have a passing acquaintance with twenty families of their own age, but they are the firm friends of one young couple who are shortly going to have a second baby.

This bond is sealed by the as yet undisclosed fact that our young family also is pregnant. They have said nothing so far, but it has not missed their notice that the other wife is going frequently to the laboratory and that their friends have had more than one family consultation recently. They broach the subject. "Yes, you should go upstairs and tell the Doctor. They see you through the whole time as you know and help you with the baby too. I only wish we had been members before we had our first. It's so easy here to get to know all you want, and things don't go wrong as they did with us. It's because you start right from the beginning, I suppose".

Later that same evening our couple are dancing again, this time not only together but with other good dancers too. They see the biologist, who having just finished the evening overhauls, is coming leisurely through the crowd talking to this one and to that couple. "I'll speak to him now, Flo". The two men talk for a while. "Yes, she's over in the cafeteria ; we will go over ; where is your wife ?" The three go over to the other side of the bath and there find the woman biologist who carried out the wife's overhaul. "Are you free ? Come into the office a minute". The four go in ; the news is broken ; congratulations follow ; everyone is pleased. On leaving the little office the pair go straight upstairs to the Receptionist and make an appointment for the wife to come to the laboratory the next day.

With the announcement of conception we are brought to one of the most critical phases in the life cycle of the family ; critical and moving for them but critical also for us as biologists because conception heralds one of the periods of phenomenal change when the family undergoes very rapid development. Like the ovum after fertilisation, by conception the parent-organism too is thrown into a fluid or plastic phase and is ready to differentiate. It is keenly alive, questing, sensitive, and in its expectancy peculiarly open to new influences.

Through their previous health overhaul this couple have gained a knowledge of their physical condition and of the presence of any disorders or deficiencies, and they have had time to put these right. Through their use of the Club they have exercised and developed their physical and social faculties and by now have probably reached a high water mark of fitness. They *want* their baby. They are conscious of taking a great stride forward at conception.

In such a setting it is no surprise to find that one of the first fruits of the Centre's work has been the preponderance not only of wanted babies—most babies are wanted after a certain stage in the pregnancy—but of *wanted pregnancies*. Apart from the emotional aspect of a conception welcomed *by both husband and wife*, its significance for the development of health is very great. In view of the subtlety of the mutual relations of a physical order, discussed in chapter II, it is clear that we could only hope to begin to see the full expression of functioning in a family acting in unison. Were the relationship between husband and wife not of a harmonious and mutualising character from the outset of pregnancy, and indeed from the early stages of their association, their later functional expression must inevitably be arrested. Any factor tending to dislocate this mutual relationship early in the pregnancy must disturb the smooth rhythm of interwoven endocrinological events, and thereby blur the functional picture for which we as biologists are looking.

In the case of the young family we have been following, both husband and wife were shy, and their tempo in taking hold of new knowledge to match their actions, somewhat measured. There are other families who grasp more quickly the significance of what they hear at their first family consultation and of what they see around them in the Centre; families who understand the significance of the fact that from the moment of conception the very tissues of the future child and the quality of its nutriment depend upon the moment to moment state of the mother, varying directly with her relation to her internal and external environment. They realise in a flash that it is of the greatest importance that both parents should be in their fullest health *before* conception occurs.

The action of such young families has been even more arresting. Quite spontaneously they have come some months beforehand asking, for instance, that their periodic health overhaul might be put forward because they were expecting to go on their holiday on such and such a date, and hoped to start their first baby then. They wanted to know that they were as fit as they could be when conception occurred. This thoughtful and express use of the consultation service arose within two years of the Centre's establishment, and it seems to us that this, perhaps more than any other happening, affords confirmation of the sure grounding of the Centre as an organisation capable of promoting health. The same intelligent use of the consultation service is also to be found in families joining after the birth of their first child, before they set out on a second.

More significant still, there are young families who before joining the Centre had determined never to have any, or any more children, but who after being members for a year or two also seek a consultation to tell us that they now intend to have a baby. They have changed their minds owing to their fuller and different understanding of the meaning of pregnancy and the ease and success with which they see it accomplished by members of the Centre.[1]

These actual happenings, pointing to a changed outlook towards pregnancy brought about by the Centre, should be of the greatest

[1] 59 out of the 91 babies born within Centre membership were conceived after their parents had joined. Of these 59 conceptions :—

- 6 were actively resented ;
- 23 were accidental but more or less readily accepted.
- 30 were deliberate.

Of these 30 wished-for babies :—

- 10 were born of parents who had previously determined to have no, or no more children, or not to have children until after some looked for but uncertain improvement in their financial position.
- 3 were born of parents whose marriages had not been consummated at the time they joined the Centre.
- 6 were born of parents who were previously infertile and who responded to treatment.

The birth of 19 children can therefore be said to be directly attributable to the Centre. Of these 19, 1 was a fourth child, 1 a third child and 2 second children, the rest being first children.

interest and encouragement to those concerned with the problem of the falling birth rate. One of the acknowledged difficulties about the birth rate is its apparently consistent fall with what is called a 'rising standard of living' and increase in 'education'. These advantages of modern life induce what appears to be a reflective mood in many young couples, leading some to prefer one child for whom a higher standard, so hardly gained, can be maintained, rather than many children open to unknown hazards. Others suffering from an access of fear, believe that their children, only to be born into industrial deadlock and war, were better unborn. In either case, their attitude would seem to imply a failure to appreciate either intuitively as in the past, or consciously—as some members of the Centre have come to do—the significance of children in the cycle of their own development. Such negation of function is a sign of *devitalisation* and such a line of argument indicative of a lack of 'culture'—in the legitimate use of that word—if ever there were one.

In this alarming and common staying of natural development which finds its expression in reaction against having children, we discern once again the mechanism of *compensation* at work. The high incidence of disorder, much of it of a cryptic nature, the greater proportion of disorder present in women than in men, and the relatively high incidence of disorder in the child-bearing period of women,[1] all of which the Centre has disclosed among its members, must represent fairly accurately conditions that are general throughout the populace. Here then, is a physical basis capable of inducing an attitude of retreat from the adventure of functioning—a retreat from *living*. Unaware of their own devitalisation, the family quite naturally rationalise their situation and attribute their reluctance to have children to one or other external cause—financial, domestic, or occupational considerations.

Although in the healthy there is power to adapt to almost any environment, an arid environment deficient in chances for action and experience at the time of courtship and early marriage may operate to stay subsequent development and cause the organism to turn in and feed upon itself, ultimately consuming its own reserves endogenously. In the field of function nothing is more clear than that man does not live by bread alone. But the

[1] See Chart, I.B. frontispiece.

more a couple is subjected to long-continued and often unrecognised want, the less facultised it will become; hence the more physiologically diseconomic become its limited actions and so the more bread—i.e. material goods—it wants and clings to. It cannot find satisfaction in what is *enough* for those whose earlier experience has led them to more accurate co-ordination and who have therefore developed a more efficient mechanism for utilisation. This only emphasises the need for an environment adequate in variety of chances for the family *at the early and critical phases of its growth and differentiation.*

Perhaps at this time which should be that of the young family's natural development through the bearing of children, we see more clearly than at any other period, the pivotal work that the Centre is doing in providing the young couple with an environment in which their physical and social capacities may progressively find expression, and also with a continuous source of information to draw upon at each stage of their unfolding relationship. With knowledge at hand their outlook is widened and they expect change and development as their relationship deepens, and so they come to look for children as the expression and confirmation of their own increasing maturity. They see life with new perspective, and begin to taste its fruits in their own growing experience. This imperceptibly but profoundly affects their actions.

2 PREGNANCY

Let us come back to the young family we have been following The next day the wife comes for her laboratory appointment. She already knows the receptionist and the Sister who is a midwife. The laboratory technique is familiar to her—in fact it is all just like her first overhaul a few months ago; weight, height and all the tests over again. As she goes out she makes her appointment to see the woman biologist, not as at an Antenatal Clinic where her attendance would immediately make public her pregnancy, where she would know no one and where the doctor would be a stranger to her, but by an appointment with her own 'doctor' as for any other consultation in the Centre. On this next visit she strips as before and puts on her examination gown. This time there is a quiet expectancy about her entry into the consulting room. The examination is a complete one as before and the woman biologist, and Sister who is also present, give special attention to the physiological details of her daily life uch as food, sleep, clothing, domestic routine, etc. As she goes out, the wife makes an appointment for their first 'parental consultation' so that the results of her examination can again be given to husband and wife *as a family*.

Through the means of the first periodic health overhaul, which took place before pregnancy was in question, all pathological obstructions to function in the family have where possible been eliminated, and the wife has already been examined with a view to her functional capacities. Moreover the pair have lived in an environment in which the physical and social equipment of the biologist have become a natural part of their daily life. Thus pregnancies occurring within membership of the Centre begin under far more favourable auspices than do the majority. From before the time of conception the biologist has been proceeding as a cultivator anticipating growth and development—not as a doctor warding off danger after pregnancy has occurred.

So, in the Centre we do not expect to find 'morning sickness', general indisposition and depression as the first 'signs' of pregnancy. These conditions, common among the general public and among those families who join the Centre after the pregnancy has begun, have become for us 'signs' of cryptic disorder or deficiency. As such they are immediately attended to if they occur, but their occurrence is the exception among those who have used their previous overhauls in preparation for the future.

In chapter II we saw how after nidation of the fertilised ovum in the wall of the womb, there springs into existence a new functional entity, 'the pregnancy'. Here mother and foetus are linked into a unity by a special organ—the placenta—newly formed for that purpose, and built into the wall of the womb by their joint operations, both of them contributing to its substance and to its construction. It is through the selective membranes of this placenta that the many and varied contents of the maternal blood and lymph reach the foetus, there to be built up into its organised substance under the unified control of 'the pregnancy'. It is through the membranes of the placenta equally, that pass the substances from the foetus which stir further developmental changes in the maternal body and at the same time lead to preparation of the food necessary to the child when born.

It is then directly from the placental blood-pools derived from the mother's blood that the growing embryo draws its nutriment throughout the pregnancy. Thus, what the biologist as cultivator of the soil looks to from now onwards is the sustenance of a *high concentration of all the woman's reserves*, so that she, as direct purveyor to 'the pregnancy', can meet the demands of the accelerated, versatile and qualitative synthesis that has been set in train.

The rise and fall of some important reserves can be followed in the laboratory. Comparison of the results obtained at previous health overhauls with those found during the pregnancy yield valuable material not usually obtainable. Let us take the iron-content of the blood as an example.

In "Biologists in Search of Material" we have shown that out of 1,660 individuals examined by us in the first eighteen months' work, 657 gave evidence of iron deficiency from the *biological* standpoint ; that is to say their Hb. value on examination proved to be less than they were capable of reaching when freed from disorders, when an additional supply of iron was given, when stimulated to lead a more active life physically, or in face of all these conditions together.[1] We are now able to give further figures derived from a total of 4,002 individuals. These substantiate our previous contention that the iron reserves of the majority are, from the biological point of view, inadequate. We would particularly call attention to the figures for women in the child-bearing period, where there is a conspicuous rise in the number of those showing a deficiency.[2] Indeed, it is far worse than this, for the degree of Hb. deficiency reached by women at this age is markedly greater than in the other sex or than in adult females at a later age. From the functional standpoint this can only mean that women, on whom the future generation depends, are not in an optimum condition when conception occurs.

In the Centre, as we have shown, a large number of women who become pregnant have already had their first periodic overhaul and have therefore had—and taken—the chance of bringing their Hb. up to what for each is a high level.

[1] Ibid, p. 53, and 71-75.

[2] For a comparative table showing the Hb deficiency, from the biological standpoint, found in men and women, see Appendix VI.

Out of 101 pregnant women studied we found that *two*[1] were able to retain their previous high level of blood-iron (90%—100%) up till term without substitution. From this we gather that it is possible for the reserves to be naturally maintained in the face of the enhanced demands made by pregnancy.

In all the others (99 out of 101), the iron content has begun to suffer a steep fall at or about the fourth month. We can give no indication as to how far this might have fallen, because in the event of any fall below 90% of Hb., every pregnant woman at the Centre has been provided with an iron or liver-iron substitute to bring her iron up to *her own* highest possible level. Thus, in face of a persistently low level of fixed reserves in her body, an attempt is made to sustain a *floating reserve* so that the developing child, who needs iron to make its own blood, shall suffer no deficiency.[2]

The question at once arises, does substitution of iron for example, in the maternal body at this time serve the purpose of mutual synthesis for 'the pregnancy', and for the future child? Our experiments indicate the probable answer to both these questions. In a few cases the iron content of the blood was maintained at about 90% as long as the liver-iron substitute continued to be given, but began to fall within 14 days if it were left off. In the majority it was not maintained at a satisfactory level; i.e. above 90% Hb., but was brought to a higher level than it would have stood at had the woman not taken the iron, for, when iron administration was stopped, the level fell still lower.

Subjective evidence from the family as to the benefit to the woman of the iron substitution is strong. Wives have "never felt better" nor ever been more active "in spite of" the pregnancy; or, they have never felt so well in any previous pregnancy—i.e. before joining the Centre. Their vigour and appearance of health, their sleep and activity bear this out in a most remarkable fashion. This is further confirmed objectively by the rise in general tone, by the maintained or raised blood-sugar values, and in the blood pressure readings as registered in the Plesch tonoscillograph. It therefore seems clear that the mother is able to take advantage

[1] It is perhaps worth mentioning that these two were sisters.

[2] This has been done with every substance a deficiency of which we could trace or suspected. As the attempt to substitute for existing deficiencies was part of an experimental investigation that the Staff of the Centre were pursuing, in this case, contrary to our usual procedure, all drugs and substitutes were provided free of charge to the pregnant family.

of supplementation during this period of unusually active turn-over. This is only what we as biologists, aware of the formative nature of this phase of pregnancy in which rapid and impending differentiation is taking place in the family, would expect to have found.

Now as to the child. There is already some factual evidence that it can benefit from substances fed to 'the pregnancy'. Among our infants born within Centre membership is a higher proportion in whom the iron-content of the blood remains at a level above 90% Hb. during the first year of life, than among those infants whose families joined the Centre only after their birth. There is practically no iron in breast milk, so that the child must lay in sufficient store during foetal life to carry it over the suckling period till it can take iron-containing table foods. Though nearly all the infants born members of the Centre started their weaning on to table foods at about the fourth month, so also did the majority of infants whose families joined within two months of the birth of the child, so that the higher percentage of iron in the blood of infants born within Centre membership does not depend on their early weaning. We must therefore infer that the difference is to be correlated with the circumstances that obtained during foetal life.

No clinician considers an infant 'normal' in whom the blood iron content is low, and it is common practice to give iron by the mouth to such infants. It is, however, recognised that post-natal iron administration has to be continually repeated because the child tends to relapse and does not seem to have the immediate capacity to retain the iron given as a drug after birth. In the light of our knowledge of function this is what we should expect. The interpretation we put upon it is that through the process of mutual synthesis carried on in a pregnancy amply supplied with iron, the floating reserves of the mother have been *familiarised* for the inexperienced foetal mechanism; whereas iron administered *after* birth to a child that has been subject to a deficiency during foetal life is inadequately dealt with because the child is not sufficiently familiar with the process of iron utilisation. The infant starved of iron as foetus has been conditioned to a low iron standard and has to learn late, without the 'physical' guidance of the parental body and therefore with difficulty, to conduct its own adequate iron metabolism.

The importance of the nurtural principle—*through the familiar to the foreign*—as Nature demonstrates it in the functional relationship of foetus and mother, is here again illustrated. The learner requires the guidance of the parental mechanism for its first lessons. We met this principle in chapter II in a general consideration of the progress of the ovum and the process of pregnancy. Here we see it borne out in practice in the progeny with sufficient emphasis to afford a guide not only for the cultivative procedure of the biologist but also for the corrective procedures of the physician.

The practical importance to the health practitioner of the foregoing strong indication that the progeny can take advantage of the floating reserves of the maternal body is great. It gives ground for an experimental attempt to produce from a stock of low physiological efficiency—i.e. poor stock—progeny of a higher physiological efficiency. If this proves possible, it means that it is not necessary to rely exclusively on selective mating among the sporadically healthy and vigorous as the geneticist (and totalitarian) would claim, and as in practice the stock breeder invariably has done. If the regressive process can be reversed by the supplementation of deficiencies at formative phases in the functional life cycle—of which pregnancy is one—it is not the closed and vicious circle it has hitherto seemed to be. It means in fact that, if we recognise Nature's impetus to differentiation as the time to choose for enriching the soil, and use Nature's instrument, the family-organism, for the familiarisation of the food fed to the new generation, then nurture will prove to be a factor as potent as inheritance.

It is possible that dietary substitution might also prove to have an enhanced effect at other formative periods in the life of the family, such as at courtship and marriage, or at formative periods in the life of the individual such as at adolescence. Time has not yet allowed us to make observations on these points. In the psychological parlour it is well recognised, for instance, that a change in the physique of an individual frequently coincides with 'falling in love', or with a satisfactory marriage. The change is attributed by the psychologist to emotional release, but it may well have a physical basis of which the emotion is but the outward expression.

In the question of raising the standard of the human stock

we must be warned against over optimism, for complete rehabilitation is likely to need several generations. This has been found to be the case in animal experimentation where, for instance, by feeding with a varied vital diet and the maintenance of adequate hygiene, it takes six or more generations to secure a disease-free rat that will maintain a standard weight, and be likely to throw a standard litter without fatality. In humans we look for more than the 'standard' animal. It is the expression of the full potentiality of the individual with his unique specificity that we seek; this can only come through the operation of parenthood bringing about the familiar nurture of the child cradled in an environment adequate both in richness and diversity.

The study of nutrition in pregnancy is one that is particularly enlightening in several aspects. A woman who has appeared well in every way at repeated overhaul, her deficiencies cloaked by a life tempered to her powers by the process of compensation, may within a week or two of conception, when a new and further demand is made on her reserves, manifest clear signs of deficiency. Or, a woman who having shown signs of a deficiency, has successfully adjusted the condition after her overhaul and has been able to maintain her apparent recovery, will often lapse with the onset of pregnancy and, with the little knowledge at present available, it may prove impossible to eliminate this by substitution during the whole course of the pregnancy as far as the *mother* is concerned; though the child always appears to benefit. Indeed many a mother in this category has remarked how different her last baby is from any of the others, not only in appearance but also in behaviour and contentment.

Owing to the close contact we have with all pregnant women and their co-operation in the use of the laboratory, we have been in a very favourable position in the Centre to observe these points. There is no doubt that adequate nutrition is no mere matter of the supply of a good and balanced diet. Unless an efficient *mechanism* is there to synthesise that food into the bodily requirements, the food itself is largely wasted. For example no amount of vitamins, no amount of calcium by whatever route given—by the mouth, injected intramuscularly or given in the form of live fresh cow's milk—will with certainty maintain an adequate calcium balance in those who manifest a calcium deficiency. Moreover, no amount of iron, vitamins, etc., fed to a woman who sits hands folded throughout the pregnancy will effect the adjustment. On the other hand, an efficient mechanism can convert what would seem an inadequate supply into sufficient to meet the requirements of the child.

For solution of the problem of adequate nutrition, it is the body's constitution or diathesis that demands study, for in nutrition *utilisation* is as important as is the quantity and the quality of the food consumed. At the present time new knowledge of the quantity and quality of foods occupies the minds of all and fires the imagination—to the exclusion of the utilisation factor. The public hence is in danger of being led to believe that the mere provision of food—free meals to nursing mothers, to infants and school children, communal feeding centres, etc., will solve the whole problem of malnutrition, operating not only as a means of national economy in time of war but as a routine mass therapeutic measure. But, as we have shown elsewhere,[1] the more disordered and the less active the individual the less able is he to take advantage of the food he eats. To feed the pregnant woman and the school child without giving them opportunity for spontaneous activity and leaving them with infestation of worms or other disorders, is both unscientific and diseconomic. It is likely in the long run to prove just one more disappointment.

The work of the Centre shows that if nutrition is to be satisfactory in pregnancy, it is no good setting about it at the third or fourth month after conception—or later. Disorders and deficiencies in the parents must be removed beforehand and, where they have been present, a close watch be kept on the whole course of the pregnancy for the first indications of reappearance during this time of maximal demand placed not only on the woman's reserves but also on the efficiency of the mechanism that has to deal with them.

When it comes to the quality of food the experimenter is faced with many difficulties. We wished to ensure a full diverse diet for certain selected families later to become pregnant, either to sustain a normal physiology or to replenish the reserves of a deficient physiology. We assumed that the ordinary market should yield good rich vital milk and vegetables. This turned out not to be so. The available milk was either not guaranteed clean and tubercle free and therefore not usable, or was only to be had at

[1] *Biologists in Search of Material*, p. 70 76.

a prohibitive price,[1] or was pasteurised in which case it no longer retained its vital characteristics. Or again, spinach, said to provide iron and hence to relieve an iron deficiency, in fact only does so in some specimens. Whether it does so or not probably depends upon the nature of the soil in which it is grown. The result of this has been that in order to control these dietetic factors we have had ourselves to establish a Home Farm to grow vegetables and produce milk—an illustration of the difficulties met with in this kind of experiment.

In view of this equivocal position with regard to food in the open market, it is easy to tell why substitutes for vital foods are now being so widely used by the public as well as in medical practice. The food available cannot be relied upon to contain the requisite vital factors for maintenance of adequate nutrition. But clearly the use of substitutes is a *therapeutic*—not a biological procedure. Indeed, it is neither rational nor practical to go through elaborate commercial processes to extract from food substances essential to life and then to feed those substances to substitute for the inadequacies of the food itself. Furthermore, it is already known that however careful the extraction process, it destroys the vital balance of the product as found in well grown fresh food; and the *balance* as between the different vitamins for example is as important as is any individual vitamin.

The human body in health is the most efficient machine, and the most economic for the extraction of essential factors from food. Disease and disorder may call for therapeutic assistance: health does not. Artificially made vitamins are essentially drugs for curing or alleviating disorder: not food for the healthy.

The practical study of nutrition in health entails some form of controlled farming directed by agricultural biologists working in co-operation with human biologists. The agricultural biologist must supply the diet known to contain the vital factors in balance; the human biologist, by studying the factors concerned in utilisation, must see that the utilisation of the food supplied is efficient. Just as in engineering where the study of oil and petrol is essential to the study of the machine at work, the chemist and the engineer work hand in glove, so

[1] Milk locally obtainable in Peckham from T.T. and Attested herds cost 8d. a pint in 1938.

the agriculturalist (and horticulturalist) with the human biologist must together study the processes of health.

During the first year of the Centre's work, a Home Farm seven miles from Peckham[1] was acquired with the purpose of providing 'vital' fresh food—milk, vegetables and fruit—for the use of the young families of the Centre. Within a year, live T.T. and attested milk from this farm was on sale in the Centre at the current price of ordinary milk in the district. A year later vegetables and fruit were beginning to come from the garden from soil already in process of being revitalised by the Indore compost method.[2] Little by little the importance of these products is beginning to be realised by the member-families; little by little their value appreciated. Both the production of vital foods and the education of the family in their significance are however long-term undertakings. For this the Centre with its social field for the growth of an enlightened tradition among its families, together with the Farm situated within half an hour's journey, and thus able to draw upon the mothers' help in the fruit picking season and on the fathers' leisure for the harvest and haymaking, forms a nucleus offering great promise and enlightenment for the future.

Let us now go with our young family to their *parental consultation* following the wife's ante-natal overhaul. Like all family consultations, this consultation is conducted by the man and woman biologist together. Unusual though it is in all but middle and upper class practice for the husband to meet the doctor who is looking after his wife in pregnancy, the presence of the man biologist at the consultation makes this appear at once a perfectly easy and natural routine. The husband is anxious and eager to hear that all is in order and to discuss the new situation into which the family has moved. He at once slips into complete acceptance of his part in the mutual experience that they are entering upon. He is often quicker than his wife to grasp the meaning of what is happening to them and to adjust their lives to the new situation.

In the Centre it is the rule for pregnancy to open favourably for the family. Thus in the first parental consultation we can usually

[1] At Bromley Common, Kent. See also Appendix II.

[2] Cf *An Agricultural Testament.* Sir Albert Howard, C.I.E. Oxford University Press. 1940.

at once pass on to the routine consideration of the findings of the wife's recent overhaul. These are given one by one, and this time compared with the findings of the previous overhauls. If, for example, the iron-content of the blood has already shown a tendency to fall, the significance of this is discussed for the pregnancy. The pair already understand from previous consultations that the iron-content is a measure of the oxygen-carrying capacity of the blood stream—transport service of the body—and that the oxygen it carries is the fuel for the body's factory. Now the 'factory' is no longer at mere maintenance turn-over but has gone into full production for the term of pregnancy. An adequate blood-iron value is thus more necessary than ever.

The growing foetus also needs iron in order to make its blood, and still more to lay in a store for the period of breast feeding when but little will be forthcoming. "What can we do about this?" the prospective father asks. "If your wife's haemoglobin is lower than it was and she takes a suitable liver-iron preparation while she is carrying, her body will use it to sustain her own needs and to supply the baby with what it requires. A form of iron which by experience we have found to be effective, can be had from the receptionist, and the laboratory is there for your wife to check up once a fortnight on its efficacy". So once again the family is given the opportunity of gathering such knowledge as is available and of acting upon it for itself.

This seeming indifference on the part of the observer as to whether information is acted on or not, arises on the one hand out of his desire to find out what the family will do with the facts if they have them, and on the other hand as a deliberate method of leaving them free to act for themselves. We have found that *the goodwill of the family concerning all its members is a factor consistently operative in the face of knowledge and understanding of any situation that concerns them.* The mere fact of themselves taking the initiative heightens their capacity to take the next step that will present itself—and in formative periods such as at marriage and during pregnancy this capability for initiative is at its highest. Hardly ever have we known any pregnant family fail to respond to opportunity arising in this way.

When all the laboratory findings have been gone through one by one, there follows a short talk recalling the way in which the baby is being built up;—not out of the food the mother eats, but from the very tissues of her body, the essence of all of which is carried round in the blood 'qualified' for the specific feeding of the growing child.

"From this we can see how important it is that the circulation of the blood should be kept in the highest working order. The essentials for this are an ample diet throughout pregnancy and an active life. To be sure of obtaining the best diet available you can order and buy from the cafeteria, vegetables, fruit and milk from the Centre's Farm specially planned to supply young families like yourselves. Contrary to what is generally supposed, in health the woman's body during pregnancy is at the very height of its functional capacity, each organ utilising its reserves, so that, far from being afraid to do things at this time, she can confidently look forward to strength and vigour *as long as her reserves are adequately maintained*".

The woman feeling the exhilaration of her pregnancy often at this point in the consultation shows a glow of pleasure. To the man it usually comes as something of a surprise. But, following on what has gone before, though strange, it has the ring of reason about it. It comes moreover from the man biologist who should "be talking sense to a fellow". So our family from the beginning is unafraid of an active life during their pregnancy.

But the consultation is not yet finished. "What about delivery ? How are you placed ? In view of all that this very remarkable process of pregnancy means to you as a family, you will naturally want to get the most out of it, and therefore to be as much together as possible throughout the whole time and directly the baby is born". Then follows a short review of the possibilities open to them for delivery so that they may discuss their plans together and with knowledge, before making their decision as to how to proceed.

In the course of the consultation one of the things that always impresses us is the gradual change in the man's outlook. Instead of his wife's pregnancy appearing to him as a process of which he can only be the shy and anxious spectator, his attitude changes to one of intelligent interest and active partnership. There is no regret for what has happened : no dread of what lies ahead This parental consultation provides them both with food for thought to be digested at their leisure. It comes at a time when their emotional urge is at its height and can carry them to new situations and actions that would be impossible to them at any other time.

From now onwards, fortnight by fortnight, the wife visits the laboratory for a repeat overhaul, and watches with keen interest the results of the tests made there. Month by month she comes to see the biologist and Sister in consultation. During these visits there is opened a wide field of topical interest over which to range—her own food, clothing and routine of life and the incidents that arise out of it; how the baby is growing and affecting her; discussion of the baby clothes she is making, preparation of their flat for the baby, etc.—many of which things later become a topic of lively discussion with her friends in the building. Continual social contact with families who have made use of this knowledge at an earlier date makes its acceptance by her easy and natural. In this way the social life of the Centre is a familiarising influence for the continuous apprehension and absorption of knowledge by all its member families.

Thus the woman grows into the new situation as the embryo grows in the womb, and gaining confidence step by step she becomes mentally as well as physically fitted to fulfil her coming task. These frequent consultations afford an opportunity for her to learn as she goes along all that is happening in the pregnancy. As she approaches labour she is anxious to know how it is accomplished, her interest in what she is going to achieve heightening as the time approaches.

Pregnancy—to culminate in labour that is no longer of dread foreboding—has become something to look forward to, something wholly to be enjoyed! It is now clear that it is not going to interfere with the couple's ordinary life but to enhance it; not to keep them more and more indoors, but rather to carry them into a wider social sphere. They come to badminton next evening as usual, the girl decides not to fall out of the 'keep fit' class, she swims and carries on right up to the week before confinement, usually choosing the women's swimming club afternoons when she goes in with twelve or more other friends. Not that she need be shy. Something remarkable has happened to her during these months. Her face has lost its hard lines, her complexion has cleared, her eyes are bright, calm and steady, she has acquired a poise both mental and physical which has changed her. She carries herself differently from most women in pregnancy. The change calls to mind a flower unfolding. It is arresting to watch her during this period.

She sits in the cafeteria after her swim knitting. There are others knitting the same sort of garments. They may already have met in the reception room upstairs at some of their visits to the laboratory or to the biologist. They talk of their experiences and compare their own with those of their sisters and acquaintances who are not members of the Centre. ". . . . I always dreaded it. I thought Fred and I would never be able to go out together once I became pregnant. Lisa's always sick and ill all the time she's carrying and her husband gets so fed up with her; and, would you believe it, for the last three months she never goes out except in her mackintosh for half an hour after dark before he goes off to the Club. But Fred's not like that, and I feel so well and alive". . . .

There are also those who were knitting the same sort of garment a month ago. They now come from the consulting room with their babies. Our young mother begins to take a peculiar and personal interest in babies—which she never did before. "How lovely this one is". She will soon have her own now, and she goes down with the young baby and its mother to see the afternoon Nursery. There she finds out something about the handling of a baby. She soon gets to know the Nursery Sister and may even volunteer to come down and help one or two afternoons a week. She will learn that way. (She did not want to learn before.) So all goes forward and when the baby comes she is to some extent already at home with the whole situation and all the possibilities it holds for her and for her family.

Her gathering experience sustains her as she moves into the future. How different from the woman who carries the burden of ignorance and progressive debility into a future for her full of fear . . .[1]

[1] *Extract from conversation in the Cafeteria*, 2nd *June*, 1939:

Mrs. X came in to tell us that Mrs. Y had had her (first) baby. The following conversation took place:

Mrs. X. "She did her ironing yesterday afternoon—at 6 o'clock was waiting to go in and by 10.30 she'd had it!"

Mrs. P. "What a difference the Centre has made to having babies. I'm not going to have another if the Centre closes".

Mrs. N. "And all of us mothers who come round here would only have had *one* if it hadn't been for the Centre and now when the baby is fourteen to fifteen months we are planning for the next. I think it's marvellous. And not minding coming in pregnant—

3 DELIVERY

Pregnancy to the biologist is no state of emergency from which the mother may (hopefully uninjured) be returned to 'normality' after the birth of the child. It is a highly active, potent, *developmental process* of the family going forward to its natural culmination in delivery. Delivery is a critical stage; yes; but critical because nicely timed by the endocrine balance of the family bringing the pregnancy to a crescendo of function at term.

That delivery has come to be regarded as a critically dangerous ordeal is no wonder, for the co-ordinations that bring about natural labour and culminate in successful delivery cannot be induced at the last minute and in the presence of fear. They result from the healthy functioning of the family from the moment of conception—or even earlier. Hence to provide for the time of courtship, to make knowledge and a fuller life available for the young married couple, to give them the chance of setting in order any physical defects and deficiencies *before* conception—all these are essential to the maternity practice of the biologist. They are the cultural procedures without which he does not expect to see natural and spontaneous delivery at the culmination of pregnancy.

It is often forgotten by doctor and layman alike that with daintion in the womb there is set in train a series of physiological events destined to call into use all the reserves of the maternal body. How great are these physiological reserves we know. Each organ is capable of some seven or eight times its average resting output, so that some seven-eighths of the body's capacity remains latent in reserve for demand, as indeed we have seen in considering the power of compensation for disorders. The extent of the reserves can be appreciated for example in the musculature of the womb, each fibre of which enlarges during the course of pregnancy to eight or ten times its non-pregnant size. This is no mere anatomical device to enable the womb to contain the growing foetus. The contractile capacity of these muscle fibres has likewise increased, as indeed it must if they are to play their part in the great event of birth.

that makes a difference too—instead of creeping out after dark—to come round here in the afternoons".

Mrs. X. "I never went out the last fortnight while I was having Beryl but with John I was here four hours before; and there's Mrs. Y in here all the afternoon the day before."

In pregnancy every organ of the body can respond in like fashion—the abdominal musculature, the heart, the kidneys, the skin, etc. These reserves of the body can in many cases be measured from observation on isolated organs in the physiological laboratory; and the clinician is also well aware of them, for instance, in the insane and the dying, where they are regarded as 'uncanny' in their power. We cannot yet measure the sum of the co-ordinated functional reserves derived from mutual synthesis of the organism in health. In full functional co-ordination they must be even more powerful than in the extremities of sickness, or than in the conditioned environment of the physiological laboratory.

Behind 'the pregnancy', then, lies a great storehouse of organismal energy and material which in health is freely available. In full production these reserves are, as we have seen, called into use, turned over, *qualitatively* changed[1] and added to by mutual synthesis. The pregnancy backed by a full reserve store is throughout gradually calling this into service. As the 'slack' in all the maternal tissues is taken up in the increasing turnover, the woman becomes conscious of an exhilaration and deepening confidence born of her gathering capacity. She knows—cannot help knowing in view of the usual attitude towards pregnancy—the dread with which 'confinement' is approached by others around her, and indeed by the medical profession, but she has a quiet inner assurance of being able to fulfil her purpose unaided and successfully. So it is with rising capacity and full confidence that the mother reaches her time, attuned to maximum effort but not tense—like an athlete in training awaiting the starting pistol.

The foetus has been developing in strength and stature *pari passu* with the mother. Thus still in mutuality, still operating through the placental organ as we saw in an earlier chapter, 'the pregnancy' is brought to term through the mutual action of mother and of foetus.[2] It is not then the mother and not the obstetrician, but *the pregnancy* that labours and is delivered.

So here, in the Centre we see a wholly unaccustomed picture of pregnancy; one of *a gradually rising flood of capacity of foetus and mother reaching full tide at the approach of labour*. It

[1] The mother is sensitised to her child, and the foetus to its mother.

[2] This is no new theory in obstetrics; but it does imply new practice.

presupposes, as we have already said, a family not only in physiological balance at the time of conception; not only functioning in full mutual synthesis during the course of the pregnancy, but also freed from disorder and deficiency before the pregnancy began, and having a full complement of reserves maintained throughout its course. It is in fact a family freed from fear and from the *causes* of fear before entering upon labour.

Nature has destined the woman to be at her best in pregnancy. Hence in biology there is no basis for relegation of the pregnant woman at the time of delivery to the position of 'patient', to which modern civilisation has condemned her. Neither is there any basis for the conversion of the natural crisis of birth into the catastrophic crisis that delivery at the present day is tacitly accepted to be—above all by the so-called 'health authorities'.

It will already have become clear to the reader that the ordinary conditions available to-day for 'confinement' are not those suitable for delivery in a functioning family. The family in full function will want to continue in the greatest intimacy as delivery approaches and immediately after it is accomplished. Like all nesting animals, they will want labour to take place in the 'nidus' or nest they have prepared, and which through long weeks has already become 'familiarised' for the infant.

The woman reaching the acme of her functional potentiality through pregnancy will sense her capacity to bring her child to birth. She will not wish for nor tolerate the dulling of her instinctive responses by anaethesia, twilight sleep, induction of labour—even Caesarian section—thereby foregoing her part in the mutual action that is to effect birth. Nor will she seek 'confinement' after delivery. Having together with the foetus accomplished labour, she will want and be ready to get up and go straight on with life. Her baby demands attention when it is born and the mother is its jealous collaborator. She will *want* to have and to handle it forthwith. And rightly too, for anyone who has watched the cow licking her new-born calf will have noticed the effect of each bout of licking on the spontaneous contraction of the uterus, hastening the delivery of the after-birth and the subsequent contraction of the womb, to say nothing of the stir and stimulus that her familiar contact must provide for the establishment of the new circulation and physiological orientation of the calf itself. The mother too will want to lose no chance of knowing the baby's responses—and her own with them—to the circumstances so rapidly changing for both of them.

Nor indeed do these desires change with the birth of the second or subsequent children. In health each successive child seems to have the same arresting and instinctive interest for the parents ". . . . Doctor, I must tell you something. When I look at and hold baby, it's just as if she were my first . . . . the wonder and the closeness of it makes me cry. I can't help it . . . .".[1] The biologist cannot deny the significance of this natural emotion in the physiology of the family organism. The obstetrician cannot deny that the unfortunate circumstances of the Maternity Hospital which separate mother from baby, and both from the father, place them and *him* at a disadvantage in the conduct *even of a pathological confinement*—for it was not mere sentiment that that mother expressed; it was modern endocrinology.

The Centre provides families, especially young families, with circumstances in which the chance of proceeding through pregnancy in very full function is present. But it cannot provide for them dwellings in which a nidus fit for the birth of the child can be suitably made. The houses from which member-families come are for the most part wholly unsuitable for the conduct of labour. There is no room for an attendant to assist the mother; if there are other children, nowhere for them to go; often no running water on the floor of their flat and only one lavatory for the whole house, shared perhaps by two or more other families —all of which conditions, in relation to the possible circumstances of modern life, are inadequate. Clearly in such circumstances delivery at home can seldom be recommended. Private Nursing Homes, were they desirable, are financially out of the question, so the Hospital becomes the only possible solution for the very large majority.

It is clear that families approaching labour as the Centre makes it possible for them to do, could not be submitted to labour in conditions in which the expectant mother is regarded and treated as a 'patient'. Some sort of special provision had to be made to mitigate the circumstances with which they would inevitably be faced. Owing to the courtesy of the staff of one of

[1] Extract from conversation with mother aged 30, two days after the birth of her second child.

the teaching hospitals in the vicinity of the Centre, we were able before the Centre opened to make arrangements whereby our mothers could be received for the period of 48 hours covering the delivery, after which they return home with their babies. Thus they are relieved of the necessity of 'confinement' as invalids among the sick for a fortnight following the birth of the baby. Although this arrangement does not meet all the functional requirements of the family, it is a very valuable compromise to have been able to make in moving towards fuller health.

The young family have decided that they will be together at the earliest possible moment after delivery, bringing their baby into the intimacy of the 'nest' for the establishment of lactation. They have chosen for the delivery the special circumstances described above and offered by us *only to member-families approaching pregnancy in a reasonable degree of health.* The wife arranges to go to Hospital for 48 hours, that is to say, for the labour and the sleep that follows it, and to return home at once with her baby. She has been absolved from the routine of the Ante-Natal Department, but has already seen the obstetrician before arriving for delivery, once when going to book up at the Hospital, and once at the 36th week on a second visit made in order that he may satisfy himself that her obstetric condition is as it should be. On this second occasion she takes with her the results of her last periodic health overhaul and the fortnightly laboratory dossier of her pregnancy from the Centre, so that the obstetrician has full knowledge of her diathesis, of her condition before conception as well as detailed information of the course of the pregnancy. Relations between the obstetrician and the biologist are those of a common understanding of this unusual position of a young family approaching labour in a full degree of function. Everyone concerned expects delivery to be satisfactory—and quick.

Forty-eight hours later, the Sister from the Centre is standing on the doorstep of our family's house to receive the ambulance in which the mother and baby arrive from Hospital. The mother is put to bed and is made comfortable, and the baby is put to the breast. There is no hurry, she knows the Sister, and important too, the Sister has known the family for some time—their physiology and their psychology—and therefore knows just what to do for the mother, and where to put in a word to set any remnant of the father's anxiety at rest. For the following week she will visit daily to look after the mother, who the next day may herself wash the baby and who with the Sister will look at and discuss the way it feeds, its stools, its sleep, its clothing, etc., in those crucially important first days of the new orientation of the family life.

About the fourth day, the baby is brought to the Centre by the grandmother, by a friend or sometimes by the father himself to have its laboratory examination and its first overhaul by the biologist. When they return the mother may be up waiting for them. She hears that the biologist at the Centre has examined the baby—and it is feeding time. She is content.

We do not suggest that the picture we have given of pregnancy applies to all family members of the Centre. Far from it. It applies to very few indeed. There were only twelve families out of the 101 pregnancies in the Centre who were in sufficiently good health, whose housing accommodation was adequate and whose circumstances otherwise made it possible for us even to offer them this provision. Of these, eight only accepted the opportunity.[1]

This picture we have given of pregnancy as seen in the Centre is in some measure a composite one. Not all those women who show an increase in capacity during pregnancy remain fully active up to delivery. Take swimming as an example : not all were swimmers to begin with, and only some of the more confident would venture to learn to swim during pregnancy. Of those who were swimmers, only three or four swam regularly up to fourteen days before delivery, and only one to within four days of this time. But where they have done so, nothing but benefit has resulted. Then again, not all families work in perfect co-ordination and are courageous. Where the wife has enterprise, the husband may be timid and so tend to exert a restraining influence on her activities, either from over concern or from social or purely conventional habituation. Courage is very largely a matter of health itself. Health—wholeness—in family function invokes adventure, but the woman's courage is dependent upon

[1] Six mothers returned home with their babies within 48 hours, one was retained for a further three days, and one remained for the full period of confinement.

her reserves, and, as we have shown, the number of women in whom the reserves are adequate is small indeed.

Many families joined the Centre only after pregnancy had begun[1] and many others were not members long enough beforehand for their reserves to be built up—even in those directions in which we have some knowledge of the deficiencies to which they are likely to be subject, and accurate knowledge of the nature, extent and means of measurement of these is still relatively meagre. Then, mere administration of the deficient substance is not always enough. As we have said, it is *utilisation* that is the important factor. If the 'habitus' of a woman does not permit of effective utilisation, either through failure to establish the necessary co-ordinations in youth (diathesis), through lack of essential complementary substances, lack of activity or what not—there will be little chance of success in an attempt made to supplement her deficiencies half-way through her pregnancy. In these families we could not hope to find the expression of function. And we must from our experience stress this point: these families untouched by the Centre before pregnancy, represent the average families in the populace. We have already seen that 90% of our members on first examination were subject to deficiency and disorder. We have no reason to believe that this differs for the rest of the country, except in the social problem groups where there is excess or poverty, in which case the condition is in some respects likely to be worse.

4 THE PUERPERIUM

The puerperium is the name given to the ten days or fortnight following labour when the womb is reassuming its non-pregnant state, and during which it is the almost universal practice to keep the 'patient' in bed—as though the puerperium were an illness and not a natural state.

Pregnancy is a mutual process engaged in by mother and foetus alike, and this mutuality of action does not cease with delivery; it is continuous into the next phase. From the biologist's point of view, therefore, the puerperium cannot be regarded or treated as a period to be devoted to the care of the mother

[1] By the end of the first two years, the ease and success with which pregnancy and the rearing of the infant are accomplished in the Centre has become known in the district, so that many families have joined *after* pregnancy supervened in order, as they thought, to enjoy these advantages.

as a *separate* entity—as is all too often the case. Its progress and satisfactory accomplishment is intimately bound up with the infant linked so closely to her not only by the breast but in a full functional sense—as we shall see in the next chapter.

But looked at merely as a physiological operation birth is an evacuation of the uterus and must be expected to follow the same principles as all other evacuatory processes of the body—for example, of the bowel, the bladder, etc. We must then expect to find preceding the birth a series of contractions of the viscus proximal to the mass to be extruded—the foetus in this case. These contractions—the labour 'pains'—ease the mass towards the outlet, which, through co-ordinated automatic innervation now simultaneously dilates, permitting gradual onward passage of the foetus. The evacuation, like all other evacuations, is then followed by general contraction of the empty viscus, first in strong rhythmic periodicity—the 'after pains'—to be followed later by general tonic contraction of the womb.

In a body the general tone of which is well sustained, that is to say, in a woman reaching her time in full activity with her reserves well engaged, these natural contractions are strong. Following the physiological law of summation of stimuli, any *general muscular movement* after evacuation enhances them, thus hastening contraction of the womb. During the puerperium these natural and necessary contractions of the womb are still further promoted and stimulated by the process of lactation.

So, when the woman comes to delivery *in the fulness of her functional capacity*, we must not only anticipate but deliberately plan both for her continued activity and for lactation, knowing that these factors contribute to the natural rebound of the womb in progressive contraction. To keep a mother, whose physiological tone is such that her involuntary responses are live and instant, reclining in bed as an invalid and isolated from her baby at this time is to invite puerperal debility and the pathological conditions that hang upon it. It is as important that a woman in full function should be up and about her business and in constant contact with her child—at least in so far as it demands service—after she has had a sleep following delivery, as that a debilitated woman whose musculature and tone are flaccid, and whose unsustained reserves are at a low ebb, should be kept in bed and every measure taken to regain what is regarded as her 'normal' condition.

This means that in the future, as health comes to be better understood, the need will arise for two different provisions for Maternity.

(1) for the 'expectant family';
(2) for the invalid wife who is pregnant.

A Maternity Service for invalid and sick mothers will always remain a necessity, but a Nation determined to renew its policy and practice after the war, whilst not omitting to make provision for the invalid, must essentially ground its *basic plan* upon the needs of the valid.

CHAPTER IX

INFANCY

1 BIRTH : A WEANING

Parenthood is a common experience and Child Welfare a well established subject, so that the reader, feeling himself on a well trodden highway of experience, will not be expecting any new point of view. But here, too, a biological approach presents us with a very different picture from that derived from physiological and clinical studies on which our knowledge of the subject has so far been based.

The biological approach is as yet a difficult one to appreciate because, living as we do in the midst of an environment full of factors that limit the expression of function, we become accustomed to regard the family acting under these limitations as 'normal'. So that when presented with a new view of the functional possibilities the reader may answer—"But this is not life as I know it". True as this is of pregnancy, it is, with a few sporadic exceptions, equally true of parenthood after the birth of the child. New circumstances, such for instance as the Centre provides, are necessary before the full function of the family can become self-evident. Man's habits and his intervention can mask or mar the expression of function according to the conditions with which he surrounds himself.

Let us approach the subject of parenthood as biologists studying function. In conception we have seen the appearance of a new focus of differentiation within the family-organism. We have seen that after nidation in the womb the foetus and mother together constitute temporarily a new entity within the family—'the pregnancy'. This pregnancy, following its own contained design, moves by ordered stages towards birth. The child, having accomplished its first close apprenticeship gently tethered to the working-bench shared with its mother in the womb, now emerges a perfected individual as far as its anatomy is concerned. The family-organism has in fact differentiated a new specific organ or 'limb'.

The old working-bench—the placenta—has been cast aside and the apprentice is now promoted to a freer use of the family workshop. Once born, the infant immediately takes over certain tasks, for example, the control of his own heart beat, circulation, respiration, digestion, endocrine and excretory apparatus. He now approaches life from a new aspect with a wider power of excursion and exploration. But still he is not left to fend for himself—to feed like the new-born amoeba direct from the environment; he is merely promoted to the next stage of a further long tutelage in mutual synthesis within the family circle.

Where the parents in mating have already become a functioning unity, the child will not only have been conceived within but will be born into the full experiential field of that parental unity: not the influence of one or other parent, whether continuously or alternately operative, but lapped in the quality of novelty arising out of their combined function as a bi-polar organism. In this sense already at birth he might be said to be *as old* as his parents, for he begins his life from where they have climbed to in their own experience of living[1] and as he grows he shares with them their growing experience. This familiar field for his specific nurtural growth is an integral part of the child's biological birth-right. Rich or meagre, what his parents yield in experience is as much and as irrevocably his inheritance as is his chromosome complement.

The infant, now anatomically perfect and surrounded by the full flavour of parental experience, is as yet wholly unfacultised. He has no discrimination—only encased in his body a perfect set of new and untried tools with which to fashion his individuality through his own progressive facultisation. All now is grist that comes to his sensorial mill. No barrier of habit, nor prejudice instigating rejection of experience such as may inhibit his elders, intervenes between him and synthesis of the world of the moment. So the child—exquisitely alive and sensitive organ of the family organism—becomes from now on like the coleoptile[2] of the plant which stirs and draws the rising sap from the roots entrenched firmly in the environmental soil. Thus through the child's instrumentation the family is led to a further awareness of and brought into closer relation with the ever-changing procession of events

[1] This is true even in a physical sense, for a modern child has been presented in embryo through the metabolic operations of its mother's body with the products of synthesis of such exotic foods as bananas from Jamaica, pineapples from Brazil—things which were outside the experience of its mother's mother. These are familiar substances to him at birth—though he yet has to learn to digest them.

[2] The 'growing point' of the shoot.

in the environment—more strange to the parents than to the child. But though the child now becomes increasingly a contributor to the family development, it is from the sap that flows continuously from the parent roots that he and they are nourished. Cut the coleoptile from the plant and the flow of sap from the roots is curtailed ; cut away the roots and the coleoptile withers and dies. The function of the organism as a whole is necessary for the development of each part in mutual synthesis with the environment ;—that is to say for health. This experiential field in which the child is cradled is then for parents and child a *mutually* eductive zone that draws each on : the parents by virtue of the new material acceptable to and so brought in by the child as it grows ; to the child by supply to it of nutriment of every sort in continuous process of being rendered specific by the parents.

So birth is not a release into 'independence' for the child. It is a *weaning* from the closely familiarised environment of the womb to one of a wider familiarity about the hearth ; and not merely a weaning for the child but a *mutual re-orientation* of the whole family. Indeed, the transition from womb to hearth or 'nest', though it entails a big physical leap is but a short functional journey, for the organisation of the nest is functionally similar to the role played in pregnancy by the placenta. It is a new 'zone of mutuality' appearing with the new orientation of the family. So, as our knowledge grows, we must expect to find that there is a specific and potent endocrinology of the nesting period.

We are familiar with such a functional zone in the very material nest of birds or earth of foxes prepared for this period of parenthood. These are inviolate zones. They are reserved strictly for the excursion of members of the family and jealously protected against the intrusion of foreign non-specific elements. It would seem that with some birds this zone of mutuality—'inner territory' as it is called by the ornithologist—extends some 1½ ft. around the actual nest, within which circle no other bird encroaches or is allowed to encroach during the nesting period—at any rate in the parent bird's presence.[1]

The human family instinctively recognises the family hearth[2] as this specific zone of mutuality prepared beforehand for the

[1] *British Birds*. F. B. Kirkman. 35,100. 1940.

[2] We use the word 'hearth' here in distinction to 'home' to which we ascribe a different and wider functional significance. See chapter XIII.

birth of the child and the time that follows immediately upon it. Evidence of this is seen in the young family's concern in making due preparations for the baby. The parents act as though it were impossible to spend too much time, thought and money on their purpose during the preceding months. This instinct for nest-building in humans is clearly present, however seriously the provisions of our modern civilization interfere with it at the crucial time of birth and in the early days of lactation.

The explanation of the failure to appreciate the importance of 'nesting' lies, we believe, in the generally accepted circumstances of present day confinement. The mother, believing that she ought, leaves her house most unwillingly some days before delivery and continues in confinement in strange surroundings for some two, three or four weeks after the child is born. During this time she may see her husband for a short quarter of an hour on the day following delivery, and subsequently twice a week more or less in public till she returns home. As for contact with the baby, in the most favourable conditions she can *look* at its cot at the foot of her bed between feeds, but, if she is not so lucky the baby, duly labelled with its name and swathed in blankets, is brought to her from another room by a nurse, and after a feed of set duration is carried off not to be seen again till the next four-hourly feed. It is indeed within our experience for a mother to return from her confinement never having seen, still less held, her baby naked, knowing nothing of its sleep and waking periods, nothing of its feeding—whether for instance it has had a supplementary feed or any medicine while away from her; and it is quite outside our experience to meet a mother who has ever—except by accident—seen the baby's stools until her return home, although it is in the stools, as we shall see, that the baby writes its record of progress in its first and most critical lessons in digestion.

In these circumstances—which are those of the majority of mothers in any large well provided modern city—*the crucial phase of the nesting process is entirely suppressed.*[1] It is not then

[1] Some idea of the trend of events that is militating against the early critical stage of 'nesting' can be gathered from figures given below indicating the place of confinement of London women during the year 1937. Birth of the child in the home occurred in less than one third of the total confinements for the year. Two thirds of the families concerned were, therefore, deprived of the biological stimulus to parental function at one of the most critical periods in the family life cycle. (*Continued on p.* 166)

surprising that the family, thus unwittingly disrupted at the time of birth when the phase of intimacy and exclusiveness is at its height, should give evidence of no more than the pale weak afterglow of a frustrated expression of the nesting instinct—which then seems to all concerned to have no particular meaning or importance. The function of parenthood has been nipped in the very bud.

But, in the Centre we find a pronounced expression of the nesting process. Though before and during pregnancy many young husbands and wives have grown accustomed to a life of sociability and action wider than is usually possible in urban society, immediately after the birth of the child we see both parents withdraw from social activity and become absorbed in their intimate concerns. In other words the pair go into a 'centripetal phase'. It is as though the family were digesting a full meal of experience; and while doing so were disinclined for other action.

During this time the family is very little seen about the Centre, though they come without shyness and with interest—often eagerness—for their consultations in the intimacy of the physiological department. Above all, the mother, or sometimes both parents where the father is free during the day time, come for the baby's examination. But they do not stay. They hurry home, wrapt as it were in contemplation of the almost unbelievable wonder of the whole process, as well as of the baby. They do not speak of this. It is how they act. The mother resists

Institutional Confinements:	
Voluntary Hospitals	20,579
Borough Maternity Homes	2,409
L.C.C. Hospitals	19,843
Private Nursing Homes	3,440
	46,271
Domiciliary Confinements:	
Borough Midwives	176
Midwives from Voluntary Hospitals and organisations	7,585
Independent Midwives	6,858
Private Doctors	5,139
	19,758
Total	66,029

Annual Report of the London County Council: Vol. III (Pt. 1) Public Health. 1937.

being parted from the baby—will hardly let anyone hold it, not even the doctor (biologist)[1] with whom she has been in constant contact. The father is loath to be long from the mother; they will not go near the Centre Nursery nor leave the baby there even though they are familiar with it and with the Nursery Sister, and may have an older child using it daily.

This picture is so closely reproduced in the writings of observers of animals in the wild, where the mother resolutely refuses to leave the nest or lair, and is closely attended upon by the male who often feeds her in the early stages of lactation, or where the pair take turns to tend the nest, that we are encouraged in accepting it as the expression of biological function, rather then due to any social reticence or the remnant of some bygone taboo.

If birth is a weaning, and weaning a process in which *the family proceeding in mutuality* is undergoing a process of re-orientation, these first days or weeks of centripetal urge take on a new and profound significance. It would appear that in the weaning from the womb of the new born at birth, an opportunity is provided for an initial immersion of the infant in the full potency of the family *quality*. Thus, before the infant novitiate wins the freedom of the family nest it is 'sensitised' to its new environment and, as it were, receives its cue for mutual synthesis of the experience it will later encounter and through which it will develop its own specific quality—its individuality. How important, then, may be this natural inturning of the family at the time of birth, in giving to the infant its own personal chart to health.

What of the parents during this time? It would seem that their indwelling during the centripetal phase is essential in effecting their own sensitisation to the new experience and in permitting their own orientation in the newly-created circumstances. It gives the other children too, particularly the youngest, time for appropriation of the baby as integral to their family and thus to themselves. What a difference this must make to the child when, for instance, the first admiring visitor arrives to see the baby! It is *his* baby they are admiring, and he has gained in dignity and importance in having acquired it; rather than having been robbed of his position as centre of the picture by

[1] It must be mentioned that this reluctance on the part of the mother can easily be overcome. The point is that it is there, and should probably be respected in health. In the practice of *medical* art it is perhaps inevitable that it should be broken down.

some foreign intruder acquired by his mother outside his knowledge and during her absence from him. For the child this completes naturally and finally his skirt-weaning from his parents, confirming for him his independence of mobility, while making easy and natural for the parents their excursion into a new field of interest in which the older child shares but in which he ceases to be the focal point of their attention. The greater the gap between children, the harder often is the dehiscence of the youngest child from too close parental care. In the coming of a further baby, after a six or ten years' interval, we have often noted the spontaneous though sorely delayed release of an older child previously locked with its parents in an unresolved psychopathological grip.

This centripetal phase of family action is then no inert period. On the contrary, just as the centripetal phase which earlier occurs with mating is a time of heightened sensibility and plasticity of the new-formed family-organism, so this period following the birth of the infant is one of the great formative periods for the family. How important it is, therefore, that there should be some mechanism for making knowledge available to the parents at this time without involving them in new, that is to say, foreign, contacts, or in any social excursions for which functionally they are not yet ready.

In a very large measure the Centre fulfils these conditions. Through the periodic family overhaul and family consultations, contact has already been made and established with the family, not only during the pregnancy but before conception. With each recurring contact the pair have had time to grow more at home, and more able to use the information available. It is natural, therefore, for a young family in the Centre to make fuller use of these opportunities for knowledge at this very time of their heightened sensibility.

Just as might be expected, one of the first visits the mother makes after the birth of the child is to the Centre to confirm for herself the information about it which she has already had from her husband, and to check up on her own post-natal condition.

Forthwith then she goes to the laboratory for her examination and makes an appointment for the next parental consultation for herself, her husband and baby. So important do we consider this time immediately following birth, that this parental consultation takes place directly the mother's and baby's first laboratory findings are available, that is to say, some weeks before the mother's final post-natal examination can take place. As in all family consultations, this consultation is with both the man and the woman biologist, who review the post-natal findings on the mother and baby and give the parents a general outline of the significance of this period

Having spoken about the 'nest' and the meaning of the intimacies it provides, the next subject of importance is that of lactation. While the infant is finding its way in this new territory of the family nest, we find it yet once again provided with its portion of familiarised food for the first stages of the adventure, for its mother's milk is made from the same blood upon which it fed in utero. Its own mother's milk is *specific* to that child. The argument sometimes used that the artificially fed child grows and develops just as well as the normally fed child serves merely to unmask our ignorance of health. Though we have as yet no means of measuring the specificity of biological processes, the evidence of clinical findings alone is sufficient to establish that there is peculiar virtue in the quality of the mother's milk. There is the baby who does not take infectious disease while fed on the milk of its own mother; the baby with a septic infection—a furunculosis or erysipelas—that no chemical or physical therapy can cure as long as it is artificially fed and which can only be saved by the milk of its own mother—or at poor second best the milk of another woman. Human milk alone allows this baby to mobilise sufficient antibodies to survive and to recover. Again, statistics show for example that during the first year breast-fed children suffer only about half the catarrhal infections of artificially fed children.

The process of lactation in the mother is evoked by the baby who now incites functioning in the measure of its proximity and vigour. It is common knowledge that though the breast has been prepared for lactation by the mutual synthesis of mother and foetus in utero, the flow of milk is sustained by the sucking of the baby. The contiguity of mother and infant is probably of all pervasive significance. We must infer that the mother's natural impulse to see, to hold, to smell, to cuddle and to kiss her baby at this time is an intrinsic part of the mechanism of functioning. In the almost universal absence of any functional setting for the family, there has as yet been no opportunity for establishing the accuracy of this inference.

Maintenance of the flow of milk is also largely assisted by the mother's muscular and circulatory tone, so that her general condition and her activity in the puerperium are of paramount import to her and to the baby alike. For the mother lactation is a powerful stimulus to her internal secretory mechanism, causing the womb to contract and her figure to regain its litheness. But it is destined to bring about not merely involution of her organs to the non-pregnant state, but her own evolution to a further stage of maturity. We do not know the time necessary for these changes, only that they depend upon the mutuality of function between her child and herself. We do not yet know how this increasing maturity affects the man, but it is unlikely that the highly specific substances produced by the mother at this time leave him unaffected. This also is a matter for future research.

In the womb the foetus was fed directly upon the placental blood with no intervening alimentary apparatus. Once born, the child must feed through its own alimentary canal, all the organs of which are formed and ready for use by the time of birth. But it cannot instantly master the use of these organs, any more than when born it can immediately use its legs to walk, or its eyes to see. It is a learner from the moment of birth just as it has been a learner in the womb, but now it is promoted to a greater degree of individual action. It must use its own lungs to breathe, its heart to direct the flow of blood for itself, its kidneys to excrete—all operations needing a long training in the acquisition of perfect co-ordination.[1] The same is true of the use of its alimentary apparatus, stomach, bowel, etc. Alimentary digestion is one of the first lessons after birth, and to this the infant's close attention is given. At first, it takes but a taste and sleeps continually till it is ready for more. For a varying number of days, two to ten or more, according to their combined sensibility, the infant and the mother are establishing the flow of milk, the composition of which is changing from day to day as the infant sucks. First a little, then more, till the flow is a steady one, during which time the mother's endocrine balance is being continuously and rapidly adjusted.

[1] No clinician lays any great stress on variations in respiration and heart-beat until childhood is well past and stabilising co-ordination has been acquired.

But the infant has still to learn the process of assimilation of milk through its alimentary apparatus. So it is that one of the first signs that the biologist looks for in the life of the new-born is its *establishment*, meaning by this that it has mastered the process of digestion of its mother's milk.

Establishment is recognised by the general aspect of the infant, which loses what is sometimes an anxious and always a pre-occupied mien and acquires a serenity. Its body fills out; its skin now fits its figure accurately; its eyes open widely when awake and like a satisfied puppy it sleeps deeply and peacefully in the intervals. It has achieved a steady rate of gain in weight, and a consistency—sometimes even a regularity—in action of the bowel.

In our experience of all types of infant, this 'establishment' takes place at any time from 2½ weeks to 3 months after birth—it may be even later in an invalid infant. At whatever time it does occur, it indicates that the child's alimentary system has established its power to utilise its mother's milk and that that food has performed the useful service of bringing about a further stage in the development of the infant's alimentary co-ordination. The mother's milk has, in fact, been for the child a gentle instrument of education. To these weeks between birth and establishment then, we might well apply the term '*appetitive phase*' for this primary alimentary co-ordination.

No value can be placed at present on the time at which establishment takes place, because, where full functioning of the family at and before this period is almost universally in suspension, even the pitch of the infant's physiological operation is likely to be queered. But keeping in mind the existing conditions of society, the great variation in the time taken for establishment in individual infants on the breast may well indicate the tempo of development of that individual (and/or of that family) for all other co-ordinations due to appear as the child grows, and so afford a hint of the likely rate of response to subsequent eductive factors in the environment and of the likely date of onset and duration of future 'appetitive phases' for other co-ordinations. This suggestion points once more to the difficulty in preparing any field for biological experiment and to the long term observations the human biologist must envisage in the study of function.

It is not in the infant alone that we see definite changes with establishment. Changes can be seen in the whole family. Whereas at birth the parents are satisfied when they know the baby is "properly formed", now after establishment they are assured that it is going to "grow properly". This seems to bring to them a slackening of tension and a sense of release. It would not be surprising, therefore, to find that the establishment of the baby corresponded to the turn of the parental tide of social action from its centripetal to a new centrifugal phase. The infant has, in fact, established its *functional locus* in the nest, just as the fertilised ovum established its locus in the womb. From henceforth the child and the parents can uninterruptedly go forward in mutuality, just as the foetus and mother, linked through the placenta, worked in dual control in the womb.

It is at this time that we begin to see the father in the Centre again, taking a serious interest in his own hobbies, rejoining his friends; and to see the mother accompanying him on his off day, watching him at his game; and now, for the first time, the baby is taken to the Nursery to sleep between feeds while the parents are in the Centre.

During the weeks that follow, the mother will continue to bring the child weekly to see the biologist.[1] She will begin to appreciate that contacts with changing temperature, with sun and wind, etc., are 'food' as important for the baby as is mother's milk and that they induce new co-ordinations of skin and circulatory system, nerve impulse, muscle tension, etc. The infant gaining steadily in weight is now extremely busy in the task of growing; its heat production is thus relatively high. If it cannot cool itself through the air-cooling apparatus of its skin—like the car by its radiator—its growth will be checked to keep its temperature within the range of physiological requirements. In the face of over-clothing—all too common at this age—this process of adjustment may go so far as to limit the child's feeding capacity, and cause it to refuse the breast or to take insufficient. This is a point too seldom recognised, with the result that the infant is taken off the breast which is thought not to suit it!

While the 'appetitive phase' for primary digestive co-ordination was still in progess and the infant had not yet learnt to

[1] See photographs 7, 8 and 9, p. 55.

utilize the available food adequately, it needed to be maintained at a steady temperature. This perhaps accounts for the time-long intuitive practice of swaddling the new born, and we see something very closely akin to it in the refusal of birds and many mammals to leave their young for the initial period after birth. But when establishment is past and the infant is free to concern itself with other co-ordinations, it is important that it should experience variations in environmental conditions of light, heat, humidity, etc. The 'shortening' to the knitted woollen garment, the fibre of which holds the heat of the body and tempers atmospheric changes as they gain access to the skin through the interstices of the fabric, finds justification as our knowledge of the infant's development grows.

From now onwards the nature of the clothing is of great importance. So likewise is the opportunity for the baby to use its neuro-muscular system. First the grosser musculature, back and neck muscles that hold the head and balance the trunk, then the larger limb musculature comes into action. The mother's desire, still present, to hold and handle the baby gives it opportunity to exercise these—to get its first 'meal' of muscle experience gained from the spatial excursion of its trunk and limbs. There is no need to fear that the baby will get spoilt by too much attention, for a *functioning* family progresses as a whole, no emphasis falling on one of its members more than on another. It is from the starved and shrivelled family, the individuals of which lack either personal or social excursion, or both, that the spoilt child emerges.

During this period, while the baby is growing steadily, has mastered its first post-natal co-ordinations and is well in every way, vaccination is done. The parents understand that the protection against smallpox is the lesser of two evils and that the immunising substance is an 'unfamiliar' substance which the baby is going to 'digest' through the medium of its skin; and thus working on the principle of one-new-thing-at-a-time, they not only know that it must be perfectly well, but that it must not be offered new experience of any other sort while vaccination is in progress. For this reason, vaccination at the Centre is done in the pause between establishment and the first steps in weaning from the breast.

But already the mother is wanting to move into a wider field and is ready to be active. She comes to the Centre early in the afternoon, goes to the 'keep fit' class and swims regularly,[1] leaving the baby in the sun in the Nursery while she is occupied.[2] Through the attraction of the baby she makes new acquaintances. People who have wanted to, but have not liked to do so before, now speak to her. Perhaps one she has never dared to talk to herself now makes the advance by admiring the baby. She goes home. She tells her husband. Yes, of course, he knows the husband, but somehow nothing had come of the men's acquaintance. Perhaps if she were there in the evening they would all meet—then it would be different. Why should she not come? She will try coming out one night, leaving the baby in the Centre's Night Nursery[3] after he has been undressed and put to sleep. She goes to badminton with her husband and there they meet the other couple and all have a very enjoyable evening. She has lost her fear. Perhaps they will become friends.

So through the baby, the whole family field expands and the family relationships extend. The mother is now ready and free to exchange notes and experiences again with her friends—she begins to let them hold the baby. In this way the baby, through its sentient body, gets its first experience of new though friendly touch, smell and many other subtle factors which so far we are not in a position to note; still less to measure. From now onwards the mother further consolidates her acquaintance with those she has newly met in the vicinity of the Nursery. All these happenings, small and insignificant as they may seem at the time, are already, by widening the parents' environment, paving the way for the next weaning that is to follow.

2 BREAST WEANING

With the baby the weeks pass quickly. The vaccination is over, and perhaps now four months old it is sitting up by itself or standing upon the knee, moving its limbs freely, regular in

[1] Swimming appears to be a natural lactogogue.

[2] Photographs 10-15, p. 56 and 57.

[3] Open from 7.30 to 10 p.m. nightly for infants in arms (under two years). The babies taken home at 6 o'clock are undressed, fed and put to sleep wrapped in a shawl till in a deep sleep, so that by 8 p.m. they can be picked up without waking, put in the pram and brought to the Night Nursery. When taken home they can be put straight into bed, without being waked. Without some such routine young parents are cut off from social life during the only hours of leisure that they can spend together.

its habits, steadily gaining in weight. It is ready for the next step—the gradual weaning from the breast on to solid food.

It is at this point that the biologist again asks for a *parental consultation.* The time is coming to use the next step in the family's progress to lay before the parents more knowledge of the principles of weaning. The consultation is a friendly talk between four people with a common interest. There is the baby's progress during the last four months to discuss, the parents' comments on their experiences to hear. By this time, it goes without saying that husband and wife are equally interested and involved in each successive stage of the family's unfolding, so that approach to the next step is easy and natural.

Although much has been written about breast-weaning and many rules laid down for its accomplishment, they all remain in the realm of dogmatic empiricism. Some say weaning should begin any time after the fourth month ; others that it should not begin till the ninth month ; while it is alleged that the most healthy race in the world continue lactation until the child is two years old ![1]

In a functioning family, the when and how of weaning from the breast presents no problem. It takes place as part of the smooth progression of the family from one natural situation to the next. The mother enjoys her widening excursion ; she enjoys her baby ; enjoys comparing notes with other mothers and is anxious to move to the next stage which her friend with a slightly older baby has reached. So, sooner or later, not because she is losing interest in it, nor like the mother who having found no growing interest clings to her baby in her boredom and holds it back, but according to her own physiological condition and that of the child, according to her sensibility and her own inherent urge, she takes the next step forward.

How then is the weaning to be brought about ? Nature, in providing breast-milk for the earlier weaning, has given the cue. It must be by the same method ; by the use of *familiarised* food for the first step. Some suitable substance from the family table furnished to the family taste ; some food that went to feed the mother as she carried and fed her baby. As it sits on its high chair or on its mother's knee, it has its first taste of gravy from the Sunday joint, broth from the Wednesday's stew, soup from

[1] cf. *The Wheel of Health* : Wrench. Daniel. 1938.

the Friday's dinner, a spoonful of custard or pudding from the family's midday meal—food in some sense already familiarised for that child by virtue of its being the accustomed food of its parents—the food the mother brought to 'the pregnancy'.[1]

The mother watches to see how the baby deals with the new experience ; what the effect on its waking time, on the quality of its sleep, etc. Above all, as she learnt to do in the first stages of breast feeding, she looks for guidance to its stool, which will tell her how it has been able to deal with the new food. Perhaps, if the first taste has been gravy from their joint, she will say on her next visit to the Centre : "Baby liked some gravy I gave him, but the next day the stool went darker. Otherwise, he is fine". She then learns that this is natural, as the liver is beginning to engage in a new operation in the use of bile. This means that the new substance is further educing the digestive processes. Step by step they go forward, small tastes first and then bolder ones, till the baby is taking enough to replace the midday breast-feed with a two-course meal from the family table.

So far so good ; protein, starches, fats have all been tried and mastered. Replacement of the two first breast feeds of the day by breakfast will quickly follow and allow the mother greater freedom ; or the six o'clock feed will be replaced by a tea-time meal, all the while the infant's growth and mien, weight, etc., affording a guide. Perhaps by six months the only breast feed is the last one at night and that the mother carries on till the steady curve of weight-increase tells her that even that need is now passing.

Weaning from the breast then is not a question of any 'system' of weaning, it does not for the healthy weanling necessitate any patent foods or the medical prescription of a 'diet'. The family table is the diet sheet, the mother's urge the prescription, the running knowledge she has been gaining her safeguard, her experience and understanding of the baby's progress in earlier stages her guide now. For her 'the proof of the pudding is in the eating', for the fact that the child can take it and thrive is the

[1] It is instructive to watch some bitches wean their pups. One day, when the time comes, the mother swallows her dinner and promptly regurgitates it before her litter, presenting them with food actually familiarised by predigestion. The first food of our nearest relatives—the monkeys—is chewed by the mother and given to the infant on the finger. Both of these natural processes we regard as examples of familiarisation of the new food.

best criterion where each family, in the measure of its health, must determine its own rate of progress.

How different from the bewilderment, fear and subjugation to specialist authority in the management of the young child so commonly seen in families the higher their financial (and often intellectual) competence.[1] In modern civilized life, where no call is made upon the responsibility of the growing girl in the daily life of the home, and where there is no chance for her to gain running experience of nurtural processes, she grows up without 'nous'. In the face of what then appears as her lack of 'natural instinct' it seems necessary to frame rules of therapy to stop the gap where instinct has worn thin. Indeed, so busy and so capable in applying therapy are we, that we have almost ceased to look for the causes that necessitate it. So we have systems of weaning, monthly charts for diet, etc., forgetting that each child is the product of its own family nurture, and must move forward according to its own specificity and at its own inherent tempo.

But is there a developmental sequence proper to the organism, and is there a *critical* phase for each stage in development? Does delay in moving on to the next phase matter? We have reason to believe that it does. Let us take as an illustration the failure to grasp the appetitive moment for breast-weaning in the presumably healthy baby kept beyond its time on nothing but the breast—often ten months or more without other food. When offered new food what happens? The child buries its head in the now all too familiar breast increasingly inadequate to satisfy its unrecognised needs, and passionately *refuses* all other types of food. Neighbours, nurses, doctors are called in to advise and to cajole in into taking food which, had it been offered earlier during the appetitive phase for this advance in digestive experience, would have been approached with an eagerness and enjoyment resulting in smooth rapid digestive co-ordination. We have old wisdom to relearn: "Now good digestion wait on appetite, and health on both".[2]

It may be objected that once breast-weaning has been achieved no difference can be seen between the smoothly weaned baby and the one in whom the process had been delayed. Perhaps no immediate difference can be discerned in the digestive functions,

[1] Vide subjugation of the family to the 'expert' Nanny.
[2] Macbeth III. IV. 38.

but a difference is evident in the functional excursion of the baby. When the opportunity to develop is not present at the moment that the child is ready to go forward, then the child goes into what can be called a *refractory phase.* During this phase it is not only the next stage in alimentary digestion that is impeded, but the next stage in digestion in any direction, so that whatever co-ordinations are laid down in the interim are blurred and uncertain. It is as though the body's attention—like a searchlight picking out the formal path of function on the sheet of life —were suddenly thrown out of focus, giving hereafter but a pale inconclusive tracing of indefinite uncertain action to be reinforced in later life by the hard lines of compensation. It would seem to matter, then, if the appetitive phase for each co-ordination were missed, for thereby progressive specificity, the hall mark of function, has not been educed in the child.

It must be remembered, too, that we know little about the laying down of diathesis that so largely determines the direction from which the individual will be open to future attack, and less about the ultimate origin of many diseases and disorders that only become manifest in later life. Long-term researches alone can give answer to these questions, but it is our contention that it is in the timely, sequential and ordered laying down of the earliest co-ordinations, in foetal life and in infancy, that we must look for the foundations of health.

By the time the baby is ready for its third meal it is already to some measure skilled in the digestion of the family table-food, and there can now be added to its diet elements that may have been missing from the family table. It is now ready for and can with advantage share in the wider, less specific experience of the Nursery tea. There, joining with the others, it sits at their own low tables and beside the older children quickly becomes skilled in the use of its table implements.[1]

From now onwards perceptible changes appear less rapidly; the infant's course is set fair for some months to come. The constant attention of both parents and biologist is relaxed. The mother, though still bringing the child regularly for overhaul, does so less frequently; from now onwards it is seen about once a month.

[1] Photograph 13, p. 56.

By now also the mother may have taken her place on the rota of mothers responsible for the preparation of the Nursery tea. The vegetables, fruit and milk—that morning's milk—comes straight from the Centre Farm for the Nursery. These are what she used to buy for herself from the Cafeteria while she was pregnant. Gradually she begins to take further notice of their quality—"Why the milk turns to junket before you can pour it into the bowl!"—hers at home takes twenty minutes or more and sometimes won't turn at all. She wonders why. It is 'live' milk and she learns something about ferment action and the vital properties of food. "Then there is a difference between 'live' milk and pasteurised milk?"—and she finds out the various reasons why milk is pasteurised. Not all of them satisfy her, for she has already begun to know something about the quality of food. Another day perhaps she learns a new recipe. Her husband would like that too; she will try it at home. So the diet of the family table may be enriched and its deficiencies gradually made up *through the baby*—coleoptile of the family.

The mother's interests branch out in other directions. Perhaps she has decided to go to next month's dance with her husband. She must remake her evening dress and there is the workroom next to the Nursery. With other mothers working there she finds just the help she needs, and learns to use the sewing machine. They have been making curtains for the theatre that afternoon .... She could make her own more simple curtains.

So she is led on by the *availability of facts as and when she is making contact with things that are of interest to her*, all the while gathering knowledge, capacity and courage, which, gradually assimilated, are transmuted by her into the substance of their home.

In this form of *individual and topical education* there is nothing to intimidate the diffident, no theory to put off the practically-minded nor to act as an escape for the theorist; only other babies a little older than hers, other mothers a little more knowledgeable, a little more capable and practised than she, all moving in an environment relatively rich in opportunity for action, where knowledge concerning topical happenings is continuously available according to the capacity of each to take and to act upon it. So, from the nest, the home grows out still further.

3 SKIRT WEANING

Thus in the Centre by the time 'skirt weaning' is approaching we begin to see families acting with growing confidence and understanding. Practice in matching their knowledge with action evokes in them courage, a deeper understanding and ability to act with spontaneity in the future in a way that appears to be prompted from an 'instinctive' or 'intuitive' source.

Time has passed quickly—the child now almost a year old is walking: uncertainly at first, but walking. He may already have been in the baby's swimming pool with his mother; very soon now he will be able to go in daily with the group of children of his own age. In the Nursery are children up to 5 years of age and their range of activity includes the gymnasium in the early afternoon before the school children come, the infants' swimming pool, the use of scooters, bicycles and skates, puzzles, letters, drawing boards and all the usual nursery equipment. It is a screened and sunny corner of a long open-air corridor which by the use of movable glass partitions can be extended and altered, and which allows of easy access to the gymnasium and the babies' bath.

By eighteen months or earlier the child slips his mother's hand at the door and runs into the Nursery. There is no hesitation. Up the 4 ft. steps he goes, down the slide—a polished plank without sides—over and over again, untiring, till in three weeks or so he can mount with speed and come down feet first or head first according to his whim. He has passed the appetitive phase, so well known to everyone, for stair climbing. It has been superseded by step and ladder climbing and the delight in sliding down the slippery plank so high for him.[1]

Or perhaps it is the bright ball nearly as high as himself that fascinates him. Day after day threading his now by no means uncertain way between the other children of the same age or older, he will bounce it, run after it, push it, hold it between his two hands, unconsciously sensing its mass, weight and resistance, learning all the time his position in his expanding world.

The older children from the Nursery go to the gym before going into the babies' pool. He stops to watch them undressing near the clothes pegs. It becomes natural to do the same, and in two or three months' time he has joined them in the big gymnasium.[2]

[1] Photograph 16, p. 58. [2] Photographs 17, 18 and 21, p. 59.

In the gymnasium, so far unfortunately equipped only with apparatus designed for older children and grown ups, the children run, climb, swing from the ropes that they can reach, roll head over heels off rolls of matting or off the top of a low buck and walk up and down 8 ft. long slides placed at a low angle. They love these and will often walk up them on hands and toes with as much enjoyment as they slide down. The younger or newer children climb the ribstalls gingerly at first and each time higher. The more advanced climb the 'window frames'—a structure 10 ft. high consisting of horizontal and vertical wooden bars spaced at 24 in. intervals and standing 2 ft. out from the wall. This the children delight to climb, then hang by their hands, sometimes for several minutes at a time swinging their legs, or occasionally resting them on the bar below, all the while with child-like detachment watching all that is going on below. The more adventurous will climb to the top of the 10 ft. frame and stay there free from interference, shouting down to the others.

Such a picture may suggest danger and confusion but in fact each child goes about its business with neither collision nor accident. An order not authoritative nor mechanical but of a functional character prevails. The child surrounded by others all engaged in their own activities, is not only unconsciously spurred on to similar types of action, but is learning the while to act himself in mutual association with his companions. In an age when the small family or the only child is the rule, the Nursery is the place where the child at an early age contacts other children of various ages—the same age and older than himself. In a big family, at the table, at play, in the daily comings and goings, there is constantly called forth, not a recognition of other people's needs or of one's own needs—both of which are forms of egotism—but a recognition of the *total situation*. This is an all embracing appreciation of the needs of each member of the family in relation to the family as a whole. Thus in the large family an *altruism*, not of a moral order but of actual physical constitution, is born.

It is in the family nest that we must look for the first dawnings of this *'physical' altruism*. As the hen astraddle her eggs will shift the outer ones in and the inner ones out, ensuring to all warmth for development, so later the young birds in the crowded nest will move with patterned order from the hot centre to the cool periphery and back again as their sensibility demands.

But it seems as though man, somehow become devoid of virility and ignorant of the implications of his devitalisation, is reducing this family environment to poverty and to monotony by stocking it with one or at most two children. Heedlessly he is creating for those one or two an environment in which they are doomed to grow in less than optimal health ; where the family circle is so small that the child never has any practice in accommodating himself to varied activities within it ; where the parents, relatively unoccupied and with their own development arrested, clear the field for the child to hold the centre of the stage, devoting themselves to it and living its life for it according to the tempo of their own meagre understanding of what is at stake for them all.

The healthy child is continually exploring the world about it with its senses, limbs and brain. In acquiring one new co-ordination after another it is continually gaining new powers, new skills and knowledge. It would seem that each appetitive phase for new co-ordination appears in orderly sequence and has a strength and duration characteristic of that particular child. Certainly in the gymnasium the succession of activities in which each child engages appears by no means fortuitous. Broadly speaking, the sequence in early years is fairly similar in all children, though as the child grows older it begins to differ considerably from child to child.

Of this *inherent sequence of development of the faculties* we as yet know little. Montessori's outstanding contribution to Education seems to us to lie in the fact that she first recognised the presence of such a sequence in children from 3 years onwards, and brilliantly designed instruments to promote the specific facultisation of certain of the finer co-ordinations of eye, ear, finger, etc., of which her colour cards, bells and primary cylinders, and devices for the early apprehension of number are examples.

The 'appetitive phases'[1] from birth to 2½—3 years, during which time the *grosser* co-ordinations are being laid down, have

[1] Dr. Montessori speaks of the 'sensitive periods' of the child's development. As far as our knowledge goes, the sensibility of the child is in continuous operation, but the impetus to action is more conspicuously present at some times than at others. It is for this reason that we have called such phases, wherever we have found them in the development of the organism or of the individual, 'appetitive phases' that is to say times when an appetite for action is declared.

not to our knowledge been studied at all until the Centre opened up the opportunity for doing so. Time has not allowed us to make any definite statement; no science of the subject has yet been developed. We only know of variations from one infant to another, from one family to another, as we have already indicated in the variations in time of 'establishment' of each baby. At present these variations can tell us little, for they represent the expression of natural variations in development *confused* with the expression of the pathological results of compensated disorder—in the child and in the family. Not until we have the opportunity of watching the behaviour of children brought up from infancy in fully functioning families all the members of which—parents as well as children have ample scope for action appropriate to each stage; and not until we have been able to devise some methods of measurement of function, can a study of the sequential development of the faculties be surely grounded. *This study of the sequential emergence of functional co-ordination after birth is in our opinion of as great an interest and importance as is the study of embryology from which is gained knowledge of the sequential development of the individual's features before birth.* Herein lies the promise of a rational basis for a future *science* of Education.

As in all biological processes, the law governing the sequence of the appetitive phases will be found inherent in the organism itself. As biologists therefore it is our business to hold ourselves responsive to the emergence of this law and pliable to its indications. But, unless the environment holds the wherewithal for development, unless the specific factor destined to educe the specific co-ordination is present during the natural appetitive phase, the process will be delayed—often indefinitely. The prevalence of 'skirt-bound' children, deprived of sun, air and natural changes in temperature, and starved of motor and sensory experience during the first years of life, is proof that as we look at the *average* child we are looking merely at a pathological expression of early facultisation.

In the Centre where the parents have during infancy been noting the emergence of one appetitive phase after another, they grow to be more alive to and expectant of the child's progress. A mother will know of her son's first visit to the gymnasium, and standing back a little so as not to distract him, she will watch from the upper window and be as interested as is the biologist to see what he will do. That evening she will tell her husband all about it, and as the weeks and months go by she will often come to the window at odd moments. Probably in a very few weeks he will go from the gym to the learners' bath, and here too his mother and her friends will come to watch. So the mother who came to the window first in fear, gradually learns to trust in her child's capacity to act for itself. She too is learning to be an observer pliant to the natural and spontaneous process of the child's development.

In the learners' bath, with water not more than 10 inches deep to begin with, the children first spend some time discovering how to walk, for besides treading against the resistance of the water they have to learn to negotiate the shallow steps leading down to it and to keep upright on a slippery bottom. Later, they discover how to sit or kneel down in the water and to get up again without falling over. Later still, they begin to float on their tummies moving along with their hands touching the bottom. Or, when the water is a little deeper, they will throw themselves into the air from the top step—about 1 ft. high—falling flat on the water face first.

A few of the very young children unaided have discovered how to float on their backs. This intriguing experience requires a thorough knowledge of the water, for they must come to know 'with their bodies' rather than with their minds that the water will support them. This knowledge has come to them through months of playing about in and with water shallow enough for them to touch the bottom with their hands. From the same experience, and from the game of blowing bubbles with their heads under water, others have found out for themselves how to swim. This achievement had its unexpected side, for it appeared first as a wholly underwater art—alarming because the under water swimmers do not at first know how to come up to the surface if they find themselves out of their depth !

So here, as in the gymnasium, there has been the chance of many a 'meal' of free movement, essays in co-ordination which have been the astonishment of onlookers, to say nothing of momentary breathless pangs of anxiety to the observer always with the Nursery children.

It is important in describing the activity of both the gymnasium and the infants' pool to recall the conditions we postulated for the educement of function and hence for the study of function ; namely, the necessity of many and various chances in the environment, and of many and various degrees of maturity in the company (chapter IV). This is true for the infant as for the adult. The provision, for instance, of a gymnasium no matter how fully equipped, for the use of an only child or for one or two children of a family would probably lead to little more than its desultory use. It would be unlikely to induce continued and progressive action. It is the presence of other children of various ages, all moving spontaneously and by their actions inventing and demonstrating new uses for each item in the environment, that gives impetus to adventure and affords the educative circumstance.

We too are continuously learning our lesson as we watch the children so early exercise their capabilities. We have found that no child *left alone in these circumstances will attempt what it cannot safely achieve.* No accident of any kind happened to any child under five years of age during the period the Centre was open. It is important to note that no un-skirt weaned child was allowed to go in the gymnasium—also that it never wanted to go. The child's own courage is indicator for it of what action is to be attempted. But where the grown-up, mother or instructress, or an older child acting as 'little mother', urges, helps, presses or cajoles, the child's natural impetus to action and to exploration is confused ; its inherent reliance upon itself is transferred to the solicitous busybody who is hanging upon its every movement. It is *then* that the accident will happen.

How clearly now we see the true significance of 'skirt weaning' as a *family* progression rather than one involving the child alone. It is the mother's resumption of her own interests, and their expansion in her growing social life, that gradually diverts her attention from exclusive focus on this child whose urgent need for increasing independence we have seen. As the child ceases to grasp his mother's skirt, she steps forward to new interests. This weaning, like those preceding it is *mutual* and involves re-orientation of the whole family.

How different is the picture where the skirt-weaning of the child has been delayed. We see it in the family who come to join the Centre with a child three or four years old and who halting and with difficulty get as far as the first family consultation. This unskirt-weaned child sits on the knee of his worried mother, fitfully burying his head in her coat and vacillating between tears and temper ; or from his maternal stronghold of defence—for she is instant to shield him—peeps furtively at the screws and handles of the big dental examination chair which he had seen going up and down in a mysteriously fascinating manner during the family's enrolment talk with the biologists. Anxiety lest the child should misbehave, together with a certain embarrassment at her lack of 'control' as she in confusion passes him over to the father, prevents the mother—often both parents—from hearing anything of the consultation to which they have looked forward. 'Shyness' prevents the child from making the excursion to the chair for which he longs. Result—the whole family leave the consultation without having heard anything. With a sense of frustration they return home, robbed of experience they could have digested, only to draw in the family belt by one more hole against the functional starvation that the hold-up in the weaning process is bringing upon them all. Often the mother quickly senses that if they come to the Centre she will inevitably have to relinquish her hold on the child. The habit is too ingrained, the strain too great ; she dreads the ordeal and they are never seen again. The prevalence of this failure in the process of nurture is one of the most common causes of families leaving the Centre. For them it has come *too late*.

There are some families with a skirt-bound child whose courage is great enough to carry them over the initial stages of anxiety, doubt and fear. Led by what they recognise to be the needs of the child, they seek for him the companionship of other children in the Nursery and finally, to their surprise, come to find themselves also involved in unlooked for experience and adventure.

But how rare is the family which, when it encounters chances, can immediately turn them to its opportunities, because its health is such that all its members—mother, father and children alike—have moved forward from one weaning to the next in smooth progression, each stage an occasion for increasing joy and interest to all. But of a *growing number* of families this is now beginning to be a true picture. Particularly is this so in the case of children born within Centre membership. A short four years' work has sufficed to show us the use of the diverse environment of the Centre to parents with whom we had come into contact *early* in their family life. This, we now know, is how the family will develop, given the chance to do so.

CHAPTER X

SCHOOLDAYS

In its infancy the child is so closely linked to its parents that it is dependent upon them for all excursion. Visits to the Centre, therefore, are of mother and/or father and child together. And since in the Centre is gathered so much that the young family can use at this time, it happens that the frequency of their visits gives the observer the regular contact that he needs to enable him to study and describe successive phases in the functional growth of the family, of which the processes of weaning from the breast and from the 'skirt' are outstanding examples. But when it comes to the child's schooldays, the observer is not so fortunate. The Centre has no school of its own, so that the school life of the child is a closed book to us. The parents, too, are shut off from this school life, so that in present-day conditions the education of the child does not represent a further phase in the functional development of the family as a whole.

If we are right in regarding the family as a functional unity, then it cannot be in conformity with biological law that there should be this sudden break in the nurture of the child still incompletely facultised. The tendency of present day education is at an ever earlier age to *supersede parental nurture* by the technique of the educational specialist—who may well not even have the basic maturity of parenthood! It is as though, while the child—'growing-tip' of the family—was developing its faculties within the home, we said—"Now at the first possible moment let us remove this young shoot and, lest it fail to grow, plant it in new soil and subject it to certain selected stimuli" But what have we done? Cut the young developing shoot off from the sustaining and familiar sap that rises from the parental roots; severed the child from the biological mechanism through which all nutriment must pass, to be rendered familiar and so readily utilisable by the young. By the initial presentation to the child of 'foreign' substances we have in fact created the conditions in which allergic manifestations are prone to arise. In pathological terms, this means that we are running the risk of inducing inflammatory processes rather than the smooth digestion that accompanies an ordered process of development. We do not suggest that the child should have only what the parents have to give him, but that all foreign substances and experiences should initially be tempered by the family mechanism. The implication of this is that the family should move in an ever-widening circle of experience *in which parents and child develop together.*

As things are, the greater part of the school-child's life is spent in a common, non-specific environment, and one from which the family is cut off. The parental lack of knowledge of and participation in all that goes on at school is apt to be complete. Delivered up at the gate by its mother, the child goes to school for a prescribed number of hours each day. There it is subjected to a routine based upon the calculated achievement of the average child and is coaxed to action within that limit. In this process the parents have no place and play no part. Many of our members for example had never seen their children swim until they joined the Centre, though for many years the older children had been going to the local baths weekly with their schools. The pride and pleasure of the mother who first sees her child swim a length are the outward expression of a human need fulfilled. And who will deny that a father adds a cubit to his dignity—if not to his stature—when in company with his friends he sees his son do a good dive, play in the band or with no dismay guide a distinguished guest to the person he has come to visit in the Centre. This pride in the parents is one of the signs of their awareness of the totality of their situation, and it is a natural stimulus to their own progressive development.

Once they have gone to school the children come to the Centre only in their leisure, and only when this coincides with the scanty broken leisure of their parents can they all use it together as a family. Nevertheless, as from now on each uses it increasingly for his own purposes, its doings are of interest to them all. The fact that the parents accept it, know about it and use it, not for the children's sake only but also for their own, makes its use by any of them part of the family life. How often have parents remarked—"The strange thing is that before we joined the Centre we never had anything to talk about at home, and now meal times are always a buzz". The Centre has become *common ground* for the family and the knowledge and experience gained there is food for them all alike.

School hours give to the children's attendance at the Centre a daily regularity. Between 4 and 6 o'clock a steady stream of one to two hundred boys and girls enter the building.[1] On Saturdays and during school holidays they arrive as soon as the Centre opens—at 2 o'clock. On week-days many of the younger children are met at the school gates by their mothers, or as the result of the neighbourliness that has grown up in the Centre, by one deputed from among a group of friends to collect all their children.

To begin with the fledglings. They no longer go to the Nursery as in their pre-school days, but come proudly upstairs into the cafeteria where they find their mothers,[2] watch them in the swimming bath or finishing a game of badminton. Perhaps a child has been told to meet her mother in the 'Medical' Department, or to go to the laboratory for a blood test, in which case the child would have her own appointment card for the receptionist. Whatever it is, the children go quietly about their immediate business and then school bags are exchanged for the towels and swimming suits that mother has brought.

All this time a steady stream of older children flows in, those from the more distant Central or Secondary Schools arriving last. This onslaught of the school children once over, their chatter in the cafeteria dies down and the whole building becomes alive with their activity. Within half an hour there can be seen twenty or thirty boys and girls in the swimming bath where no diving board is too high for the more adventurous, while down in the learners' bath five or six little girls are overcoming their own timidity in their eagerness to swim its length and thus qualify for unconditional use of the 'big bath'. The gymnasium has taken on the appearance of a lively monkey cage. Outside on the arena the graceful and accomplished roller skaters and the young cyclists of five and six years old are threading their way among the slower moving precariously balanced learners. Upstairs in the Games Room, two boys are having a game of billiards on the small table and three sets of table tennis are in progress; elsewhere a group of girls in their tap-dancing shoes eagerly watch for the arrival of their teacher—a young married woman member. Later on, in the cafeteria and between the pillars in the main hall are heads bent over draughts, and

[1] See Appendix X (ii)   [2] Photograph 22, p. 60

chess and ludo boards, over jig-saw puzzles and books, over knitting and pencil and paper, over a typewriter; while seriously contemplating the selection of food on the cafeteria counter is a little group of three brothers and a sister, the eldest holding tightly in her hand the money for their tea. They are anxiously trying to expend their pennies in a way that will satisfy both their appetites and their tastes.

Let us follow a group of mothers in the Centre and see how without actual participation or even maybe exact knowledge, the children's activities are in general within their ken. They have finished a cup of tea in the cafeteria and they go round to the sunny main hall where they can look down into the gymnasium and watch their children.[1] It is the conventional well-equipped gymnasium, very lofty and light, with a sprung cork floor,[2] with ropes, rope ladders, horizontal bars, a 'window frame', ribstalls, booms, horse, mats, etc., and everyone in it barefooted. But here the resemblance to an ordinary gymnasium ends. A boy of 11 leaps through the air from a swinging rope and lands on the ribstalls; three boys are sitting contentedly on the top rungs of the rope ladders, five girls are playing a game on the 'window frame', while three girls and two boys have a large light ball and dodge among their fellows as they play; two groups of boys are wrestling on the mats; one boy is using the punch ball, and five small boys leap from horse to swinging rope and back again.

One of the mothers watching from above is getting worried about her little girl who is down there standing aside rather timidly. It is the first time she has been in and the mother is afraid she will get hurt. "She'll get used to it—don't you worry", says one of her acquaintances as the first mother pauses in the act of knocking the glass to beckon her out. She sees one of the staff come in, watches as he says a word to her little girl and notes that he sends out two boys who have not taken off their shoes. The new mother notices, too, that no one seems to be nervous. So, "Perhaps it will be all right", she says to herself and leaves her little girl alone.

Meantime the observer has strolled up and is standing behind the group of mothers. He notices that they are aware of and

[1] Photographs 24 and 25, p. 61.
[2] See note on construction of this floor. Appendix I.

enter into what their children are doing and show no inclination to interfere with their activities. He notes also that the timid mother is being infected by the faith and experience of her more confident friends. In the gymnasium itself he sees many figures, boys and girls moving in every direction at varying speeds, swinging on ropes suspended from the ceiling, running after balls and each other, climbing, sliding, jumping—all this activity proceeding without bumps or crashes, each child moving with unerring accuracy according to its own subjective purpose, without collision, deliberate avoidance or retreat.

Let us study this hub of activity from the point of view of a child who goes into it. He goes in and learns unaided to swing and to climb, to balance, to leap. As he does all these things he is acquiring facility in the use of his body. The boy who swings from rope to horse, leaping back again to the swinging rope, is learning by his eyes, muscles, joints and by every sense organ he has, to judge, to estimate, to *know*. The other twenty-nine boys and girls in the gymnasium are all as active as he, some of them in his immediate vicinity. But as he swings he does not *avoid*. He swings *where there is space*—a very important distinction—and in so doing he threads his way among his twenty-nine fellows. Using all his faculties, he is aware of the total situation in that gymnasium—of his own swinging and of his fellows' actions. He does not shout to the others to stop, to wait or to move from him—not that there is silence, for running conversations across the hall are kept up as he speeds through the air.

But this 'education' in the live use of all his senses can only come if his twenty-nine fellows are also free and active. If the room were cleared and twenty-nine boys sat at the side silent while he swung, we should in effect be saying to him—to his legs, body, eyes—"You give all your attention to swinging; we'll keep the rest of the world away"—in fact—"Be as egotisitical as you like". By so reducing the diversity in the environment we should be preventing his learning to apprehend and to move in a complex situation. We should in effect be saying—"Only this and this do; you can't be expected to do more"—. Is it any wonder that he comes to behave as though it is all he *can* do? By the existing methods of teaching we are in fact inducing the child's *inco-ordination* in society.[1]

[1] We cannot be surprised that in a society that has been educated on those lines we have an insoluble road accident problem.

Let us look more closely at the significance of this picture where in the Centre gymnasium—as throughout the Centre—the children proceed without any supervision or direction to use all the available apparatus; wherein each child is *a part of the whole*; where the individual without clash can thread himself through the complications of a total situation; where mutual action is undertaken in awareness of a complex situation, that situation itself forever changing.[1]

This Centre 'gym' affords a concrete piece of evidence that *spontaneity* is no quality of haphazard action necessarily leading to confusion, but is an expression of function.

That such a scene can unfold itself before our eyes is, too, a promise that given the necessary circumstances, function will forthcome; that a situation is possible in which a diversity of individual specific actions may result in a harmonious whole. It is to action of this order arising out of the capacity of unintimidated human beings facultised to respond to the total situation, that earlier we have given the name 'physical altruism'.[2]

This type of action in mutuality with the environment, noticed on many occasions and in a great number of activities in the Centre, is in manifest contrast to the egotism invoked by conventional instruction. It is the very antithesis of the action that results from training, yet training has come to be accepted as synonymous with 'education'. Training, by whatever system, can only create co-ordinations for *special purposes* by an objective conditioning of certain reflexes. This may in given circumstances enhance physiological efficiency, but it is not conducive to functional efficiency. Indeed, in our opinion, training, in inverse proportion to the age of the individual, is a menace to the evolution of biological function. Where the spontaneous emergence of ordered facultisation has been prevented, or where facultisation has failed to occur for want of timely opportunity, training is a useful remedy—a therapy in fact—for dealing with the consequent inco-ordinations. As a remedial measure it may in certain circumstances be a necessity, but it is a necessity to be deplored because, as such, it indicates the presence in society of undeveloped capacity—arising from deficiencies of a biological nature. It behoves us to breed individuals without deficiencies rather than to remain

[1] We cannot here resist pointing out the obvious value of this kind of 'awareness' in any situation that calls for man's intrepidity and skill; e.g. the fighter pilot.

[2] See p. 181.

content with a policy of supplementation through 'training,' or other means.

Functional efficiency has to be acquired in infancy and childhood, the facultisation for it occurring in due rhythm and sequence as development proceeds. As the existing social and educational systems tend progressively to deprive children at an increasingly young age of their biological inheritance, it is no wonder that training is more and more coming to be considered a necessity for youth—though not, of course, by its advocates recognised as *therapy*.

We had in the Centre one interesting example of the inhibiting effect that training may induce. Some of the children who spent a good deal of time diving and who were deemed very promising material, were enthusiastically and methodically taught by a professional—a trainer of competitors for the Olympic Games. He was an extremely good teacher and evidently an inspiring one, as the children rushed to learn with him. But what happened to those whose enthusiasm carried them through a strenuous course? As soon as their teacher stopped coming they stopped diving, and some of them never took it up or dived again with any enthusiasm. It was as though, trained beyond their natural capacity—to a pitch that was *the trainer's* standard, not theirs—their own urge was satiated and destroyed. It is true that from an objective and external standard of diving they had become better divers than had they been left on their own, but it was at the expense of their natural interest and appetite. The acquisition of 'style' cost them their zest and spontaneous enjoyment of diving.

We do not wish to imply that there is no place for training at any age. When the basic facultisation of the individual has been established, and when at or after adolescence he determines the direction of his future specialisation, he will probably *himself* embrace a course of training to perfect his skill. It is essential that at this stage instruction should be available.

But let us return to the school children. How do we decide what material to give them? One important point is that there must be available to the child the instruments in common use in the society in which he is born. For the present-day urban child, this implies that there must be at hand such things for example as bicycles, typewriters, sewing machines, wireless sets, etc., etc. A child growing up for instance in a fishing village would be ill served if the boats and tackle, however jealously guarded, were not to some degree available to him. No doubt his balancing would be learned on a choppy sea. Not because he can learn to balance better on a boat than on a tight rope, but because boats are the instruments in common use in that child's society—the instruments with a flavour of home. Unless the child has access to them, either he is not learning to balance, or he is developing out of contact with his living world, so that he cannot utilise his experience in mutual synthesis. That is the underlying educative principle that emerges from the study of function—and it applies equally to all instruments and to all knowledge. The child is born into the *zone of its parental experience.* Its facultisation proceeds by the use of those things that are pre-digested by the parents as in the course of living their own lives they fulfil their spontaneous function of nurture. These form the bridges that will bring it, as an adult, to apprehension of an unfamiliar world.

Let us look further at the list of instruments in the Centre in use by adults and children. Many are in the category of what can be called 'self-evident', i.e., the use to which they can be put is implicit in their structure, like a bat and a ball. It is only necessary to exhibit this type of apparatus for the children to seize upon it. Our observations suggest that the instruments which are most attractive are those involving individual skill. Any 'competition' is then between the *child and himself* conforming with the various natural laws of dynamics. The fall through the air in a dive controlled by the law of gravity and the dynamics of rotation, the floating on the surface of the water, the rolling on skates around curves, the dancing to rhythm, the balance on a bicycle—in all these the child is learning to move in compliance with a different dynamic. A striking illustration of an instrument of this type introduced into the Centre was a sprung canvas 'trampolin', adapted from the fair ground, for which the children proceeded to invent innumerable uses. On this the child must co-ordinate himself with the laws of gravity and the elasticity of the spring while he bounces in every variety of posture and turn.[1] It is incumbent on the staff of the Centre, working as research students, to multiply the scope and variety

[1] Photograph 27, p. 62.

of instruments of this description which carry their own incentive to action.

There is another type of instrument, of which the use is not self-evident. This includes all those tools designed or used for specific purposes. Here the user is an essential factor without which the instrument cannot make its appeal to the uninitiated. The instructional possibility of this type lies then not in the tool, nor in the user of the tool, but in the two together in action. Examples of this type of activity range from writing and arithmetic to the game or craft of most complicated skill—cricket, chess or music. It is *action* that counts ; no form of theory can be self-evident to any child.

In the Centre all instruments become self-evident to the children as the older or more adept individuals make use of them. In this environment there is no need for direction of the child's attention. The intrinsic appeal of the instrument itself, or of other people doing things, invokes the child's selective action. Instead of looking to some older person—parent or teacher—to tell him what to do next, the child learns by his own stirrings to do those things that will seriatim bring about his facultation ; learns, too, to take the first steps in the building up of his own initiative.[1] In the Centre an adult does not play a game of billiards in order to teach the child how to play, nor does he demonstrate the use of a drum. No ; the game will be going on, the band will be playing, because the participators want to play. It is the child coming to watch who transforms the players into his instructors. So it comes about that the *society of the Centre* becomes the instructor, not by intention, but spontaneously and inevitably through the very nature of the situation, for out of the abundance and variety of social action the child is fed and filled with experience. If the society in which he is moving be sufficiently versatile and skilful, it is likely to provide most of the pre-adolescent stimuli that the child needs, to be used by him in his own time and at his own individual rate for growth and development.[2]

It may well be wondered how dispersal of some 250 children to the use of all these activities comes about without regimentation, and with the assistance of no more than two or at most three

[1]Photograph 28, p. 62. Note the small boy unconsciously absorbing a knowledge of the game of draughts.

[2] cf. Significance of the appetitive phases in education. Chapter IX, p. 182, 183.

members of the staff. In the first place the equipment, so attractive to the child, is distributed throughout the building, each item set out in its appropriate place. This, together with the fact that much of it is at the same time in use by adults, encourages orderly dispersal from the moment of the child's entry.

To gain admittance to the bath-chamber or gymnasium or to secure the use of any instrument—e.g. skates or balls or books —the child must first obtain a ticket, and to do this he has to find the Curator, who carries them as she moves about the building.[1] The act of obtaining a ticket from the Curator wherever he or she may happen to be, does away with all departmental confinement of the children. However sequestered the apparatus may have to be, the Curator is never shut away with it, thus the child only leaves the general society when the use of a particular piece of equipment makes it necessary. So the children are constantly circulating freely amongst the adult society which presents them with an orderly framework in which they meet not chaos, but effort, taste, selection, skill and all the other attributes of a mixed community in action.

Having obtained the ticket the child writes on it his name, what he is going to do, and at what time. He then goes off to his chosen activity. Entry into the swimming bath, gymnasium, cricket practice nets, etc., is through a control gate where he drops the ticket into a box placed conveniently for the purpose. All equipment that is used out of doors is kept in a special room adjacent to the covered playground. Special compartments of suitable size and shape have been designed to house each article—the 36 pairs of booted roller skates of different sizes stand ranged in their appropriate pigeon holes, each hole labelled with the size of the boot. The child, on admission to the ground floor room, finds his own size skates, takes them and deposits the ticket and his own shoes in the vacant hole. When each child finishes skating he takes his own shoes out of the pigeon hole, returns the skates and puts the ticket in the box placed for the purpose. Each bicycle, scooter, tennis racquet, skipping rope, ball, bat, etc., and upstairs each pencil, badminton racquet, book, jig-saw puzzle, game of draughts, chess, or dominoes, each billiard cue and ball, sewing materials, etc., etc., is in its own properly designed niche. These 'cabinets' in which

[1] Photograph 23, p. 60.

all the small movable apparatus is exhibited are lightly constructed and can be wheeled to any part of the building or withdrawn from circulation as the Curator wishes. By this simple device of suitably housing each piece of the equipment, none of it, when taken out or returned has to be handled by any member of the staff.

All this time the children's tickets are accumulating in the boxes. Their real significance has not yet been made clear. The ticket is very important to the child, for only with it is he able to obtain access to or use of any instrument. It is important to the Curator, for it is a deliberate method of ensuring recurring contact with each child and of learning where he or she will be and what doing in the Centre. And finally, the tickets collected at the end of the day are evidence used by the Curator of the activities of each child, to be entered in the family records. The ticket thus is a common symbol current in the daily business of child and Curator alike, and equally essential to both for their business in the Centre.

Such an arrangement has also its practical and economic aspect. Let the reader picture to himself the situation every day from 4—7 p.m. with two to three hundred children moving freely in the building, each exhausting his capacity for action in as many ways as are made available, under the enquiring eye of the Curator assisted only by a student, and whose primary interest is in watching what is going on. Not only is this a method suited to the needs of our experiment, but, as we have clearly demonstrated, it is an economical and severely practical method of social organisation.

The tickets have another use—an incidental one. The child who has not yet learned to write finds a mother or other grown-up or older child to sign its card, but, once even the initials of the name have been mastered, no one is asked. The student working on the records has thus to keep up-to-date with the progress made by the younger children, so that when "MB" appears for the first time in a bundle of 500 or more tickets, it is recognised as Maureen Brown's first use of the newly acquired art of writing. The need to write on the ticket to get what she wants makes writing a very desirable achievement, especially as she is mixing with a number of older children who can already write. Here are circumstances in which the 'appetitive phase' for writing becomes self-declamatory.

As a result of this organisation of the equipment an interesting fact has emerged. We have found that without request from us or comment on the part of the child, it is customary for him to return apparatus to the proper place after he has used it. The child is quick to respond to a mutually sustained order in society. It is most intriguing to watch the newly-joined spoilt child gradually replace his egotistical expression of greedy use and careless throwing aside of one thing after another, by a sustained use of the chosen instrument completed by its orderly replacement.

It might be expected that such a large number of children let loose in a spacious building would behave in so unruly a fashion as to be an unendurable nuisance, and that left free to choose from among so many attractive pursuits they might run wildly and rapidly from one activity to another. For the first few months while new families were pouring in to claim membership, there was hysteria of this type among the children. The order that subsequently emerged was the result of the satisfaction of the physical appetites of the children, in a building peopled with grown-ups whose use of the same instruments was an enticement to the children's own achievement—not to lawlessness. It is obvious from the above experience that any education depending upon the existence of a social forum such as the Centre, must until that forum is built up with sufficient diversity of action, suffer a confusion—'growing pains' as painful to the staff as they are bewildering to the rest of the adult society!

We did not begin with so small a staff when the Centre opened; indeed individuals of many varieties of competence were introduced into the children's life in an endeavour to engage them in varying activities. Such efforts were however only partially successful, in almost every instance the success seeming to depend on the children's appreciation of the particular skill and enthusiasm of the individual, rather than on his gifts as a teacher. As the adult members came to take up their own activities—badminton, diving, the band, billiards, etc.—these activities began to gain the attention and focus the endeavour of the children, and as time went on they came to need less and less direct contact with any adult whose time was entirely devoted to them. By 1939, apart from the Curator and her students, there was for the school-children one part-time swimming instructress, the greater part of whose time even so was devoted to teaching the older women swimming, and besides this only the occasional presence in the gymnasium of a professional gymnast who was an enthusiast for 'free gym'. At the same time the Centre always welcomed any enthusiast who cared to come and use its facilities for the furtherance of his own skill, and who was at the same time willing to assist any of the children showing interest and aptitude, to acquire that skill.

We have reason to believe that some of the restlessness, hysteria and inability to concentrate, so clearly marked in the children of many families when they first joined the Centre, was due to their unsatisfactory physical condition. To illustrate this, the high percentage of cases with worm infestation, confirmed by microscopic examination of the stools, must be mentioned. In almost all cases, this was accompanied by an iron deficiency and very frequently by avitaminosis, as well as by nail-biting and other nervous tics and anti-social habits. The elimination of the infestation, together with supplementation of the known deficiencies, so frequently coincided with a marked change in the behaviour of the child that we feel compelled to correlate the two. Vice versa, re-infestation would often be discovered through observation of social action; not through the cessation of action but by the changed 'action pattern' or tracking of the child as it moved in the Centre.[1]

Apart from this, we must remember that when a child goes to school it is subjected to an educational drive that emphasises *mental* achievement and gives quite inadequate opportunity for expression of the *physical* exuberance natural to any young animal. All too often the school curriculum offers only the most meagre physical outlet, and is entirely dislocated from social life. The system has not the fluidity of a living organisation and thus does not allow of the operation of the child's own growing power of discrimination and volition in all that he does. Our experience has already sufficed to show us that where from an early age onwards adequate opportunity is provided for spontaneous physical excursion, the necessity for 'discipline', by which this excursion is usually replaced in school, becomes superfluous. Discipline is inherent in any child seeking its own adventure within the framework of a familiar and 'organised'[2] society.

[1] For the numbers of individuals in different age groups found to be subject to worms, see Appendix VI.

[2] The word 'organised' is used here in its physiological sense as in the organization of tissues in the living body.

Indeed, one of the most striking experiences in the Centre has been the ease with which it has been possible to distinguish between the high untamed spirits of health and the hysteria of repression.

All observers have been astonished at the untiringness of the children moving freely in their chosen occupations. Many who on Saturdays or in the holidays come to the Centre at 2 p.m. and leave at 6 or 7 p.m. spend the whole time in one activity after another without rest or pause even for tea. A boy of 5½ still unable to swim was seen to dive from the spring-board into 10 feet of water twenty and more times in half-an-hour. And that not just in a frenzy, but day after day, with great purpose in response to his own subjective urge to master the dive according to his capacity, content to rely each time on some struggling effort to bring him to the side of the bath.[1] Or we could cite a boy of under 4 years old who spent four hours, day after day without a break, on a pair of roller skates, till he had achieved that particular balance. The records compiled from the children's cards show that these rather outstanding examples could be matched by hundreds of others showing great constancy of effort—which is indeed the rule and not the exception in the Centre children.[2]

This concentration and perseverance that we witness as a *general* characteristic of all the children, we judge to be due to the fact that, free to select the activity that appeals to him, the child is unconsciously choosing the one suitable to the *appetitive phase* through which he is passing and, having so chosen, gives without stint his whole attention and effort until his potential capacity for that particular co-ordination is fully resolved into achievement. Another general characteristic of such sustained spontaneous action that may seem strange from a physiological point of view, is that it is not accompanied by signs of exhaustion or fatigue.

We did not in the beginning plan the children's activities on the basis we have just described but attempted to form groups

[1] This apparently hazardous exploit was an experiment by one of the staff who gave way to entreaties by the boy to be allowed to dive off the spring-board on the strength of his prowess in diving into the shallow end. The observer merely sat on the side with his legs in the water, so that the boy might pull himself ashore when he came to the surface.

[2] See Appendix VII. *A specimen record of a child's spontaneous activities in the Centre over a period of 11 months.*

and draw up time-tables and made schemes for the regular distribution of the children to activities of their choosing. All this the children very largely ignored, for it must be remembered that they were free to do so. Observation led us very quickly to abandon such methods and to evolve the present scheme by which the child can move in spontaneity of action within the general social framework.

By the end of the third year some parents, recognising the fruitful results of the methods we have described for their young children growing up in the Centre, began to ask if it would be possible to make arrangements to use the building, at present lying idle in the mornings, for the schooling of the children ready to leave the Nursery. For this, they were ready to make a financial contribution. The development of a primary school in the Centre had been foreseen by us from the beginning, but it was a surprise that the suggestion should come from the parents —and at so early a date. Willingness on the part of the parents to co-operate was the first intimation to us that the moment had come when we could begin to approach Education *not as a process directed merely to the child, but one essentially involving the family*—particularly the mother, whom we proposed to incorporate in the educational scheme.[1] Later, we shall see (chap. XIV) that one of the misfortunes of present day urban life is the lack of occupation and interests of the young married woman. But once she comes to sense the biological significance of the nurtural process, she acquires the necessary incentive to take her part in all that the child is doing.

Nature has determined that the child shall be nurtured in the social milieu of its parents. To remove from the child's environment all traces of adult influence, and as far as possible of adult action, as has been attempted in some of the modern educational experiments ; or for the adult, whether parent or teacher, to stay his spontaneous and discretionate action for the presumed good of the child, is to deny mutual synthesis and to withhold from the child the possibility of natural growth and development of his facultisation and discrimination.

Just as the little boy's trousers must be made to fit him, so also it is necessary that many of the instruments of social life

[1] See Appendix VIII. *Plans for the first steps in an educational experiment.*

should be made to his size, so that they can be used by him for their proper purpose. A full size bicycle, for example, cannot be ridden by a child, nor can he at first use a full-sized billiard table.[1] The making of smaller working models for the children has always been a leisure occupation of fathers and it is no doubt one of the natural ways, within the family, of linking childhood and maturity. There is a good deal of talk these days of a children's world, but let us make no mistake about it, the child has no wish to be relegated to a world of its own. The world of its parents, of the grown-ups, is a place of mystery and enticement to it, and as it grows it longs to share in it more and more. The processing of such sharing is part of the essential technique of nurture, and demands immediate experimental research in methods of fitting the appurtenances of the modern world to the capabilities of the child. The world of our great grandfathers was a much simpler proposition.

The contributions to be made in the course of family development are not one-sided but mutual. While the parents exert their ingenuity in the nurture of their child, the child in his activities is making contributions to the growth and differentiation of the parents. His eager unspoilt appetite for all that he encounters is one of the avenues of impact of the outside world on the whole family, for he brings within the parental circle material that without him would not come to the notice of the older members of the family, or of which they might otherwise fight shy. Certainly, many families were first led over the threshold to join the Centre by their young son of 9 or 10, or slightly older daughter, who would not take "No" for an answer; and many mothers and fathers have found themselves in the swimming bath or on the badminton floor led by the same fresh outlook of unintimated at-homeness of the child. Thus in the healthy family the parents, through this mutual action of old and young within the family, may find themselves in keeping with their times even though they are long past middle age. It is well known—as parents say—"the children keep you young".

Our experience with every type of family has, of course, also given us a clear insight into the reactions of the pathological

[1] We were obliged to seek bicycles in France and skates in America small enough for the use of the youngest of our children capable and eager to use these instruments.

family ; where, for instance, the parents are a restricting factor to the child and the activities of the child an anxiety or a nuisance to the parents ; and where it is common to find that what the child brings to the home is received with prejudice, suspicion or indifference. But in a functioning family the material that is taken into the home is never rejected out of hand ; it is looked at, and the specific discriminative powers of the family are brought to bear upon it, with the result that it may become material for growth for all, according to their needs. By the parents' discriminative action the child, too, unconsciously is learning something of the quality of discrimination.

So two facts of great importance emerge. Society and the child in the Centre are in mutual relationship to each other. The grown-ups, *going on with their own business*, continually enlarge the field of family excursion, and the child shares this continuous expansion and makes its own contribution to it. In this situation the child is never lifted into the egotistical position of being the focus of attention—of either parent or instructor. He is on the fringe of a potent zone of activity to which he is carried by the parental growth and to which he is drawn by a dawning interest. And because he is free to move in this body of society, he moves spontaneously according to the appetitive phase through which he is passing, to the particular activity appropriate to his own development. Penetrating widely and deeply into such a society, as time goes on the child may well encounter every degree and variety of skill. All these people that he knows—his parents and their friends and acquaintances, his elder brothers and sisters and their contemporaries—become naturally and inevitably his self-constituted demonstrators and instructors.

In an earlier chapter we have described the placenta as the zone of mutuality by means of which parent and child alike are led each to the next stage of development, the child the while being continuously nourished upon familiar and specific products of the 'family' synthesis. At birth the placenta discarded, a new zone of mutuality—the 'nest'—replaces it. As the child grows to the fledgling stage still another placental zone must be built where the parents, completely at their ease, can mix with their fellows in society. So the home grows to include some bit of society, experience of which the parents are themselves in process of digesting and to the flavour of which the child is thereby introduced. The Centre is just such a society into which the family penetrates and can lodge itself for its home-making, where it can establish for itself a new social 'placenta', the anatomy of which is wrought of the friendships developed, through which all are nourished.

Towards the end of the child's schooldays, it is thus in the sphere of social relationships that family tutelage is most subtle and most strong—subtle because the child and often the parents too, are unaware of it. Here in the Centre in many different kinds of situation they find rich opportunity for social adventure.

Apart from Bank Holidays and Saturdays, most of the fathers are unable to get to the Centre before 8 p.m., and it is a disappointment to many of them that they are able to share in so little of the week-day leisure of their children. But where possible, families often foregather in the Centre. For example, on a certain Friday two mothers came into the cafeteria about 3 p.m. One went upstairs to complete a health overhaul and at 4 o'clock both had tea with two other members. At 4.45 p.m. their children came in, two girls and a boy, aged 11, 10 and 12, all with swimming suits. One wondered why at 6 o'clock their towels were still dry and the girls quietly doing their homework, but soon the answer became clear. At 6.30 one husband came; at 6.45 the other. After a few words with his wife, one husband went round to the children's billiard table for his boy and then the two men bought the four 3d. adult swimming tickets. The children got their tickets from the Curator, and as the bell rang at 7 p.m. the seven leapt into the bath. By 8 o'clock the two families were dressed and having supper together in the cafeteria—a gay, contented group. Then one of the boys fetched the chessmen and had a game with his father, until at 9 p.m. the children went home and the two men waited their turn at the billiard table. Tickets were purchased for all for the play next Thursday, in which they were particularly interested, because in it young Mrs. Jones, a newly joined member introduced by themselves, was to have a small part.

Whilst the children of this little group had been giving attention to their supper, they were enveloped in the social flavour of the family; in the warm responsiveness of the parents to their friends, in the manner in which mother or father handled each situation as it arose—subtle differences on the approach of friend, acquaintance, stranger, and in the way they received the children's own friends. All the while, unknowingly, each child was accumulating a small knapsack of familiar social experience with which later to start out at adolescence into the great world.

From their participation in this social life, the children are already showing a social competence that matches their physical stalwartness and intrepidity. Visitors remark on how responsive and forthright they find the Centre children. They are able to deal with strangers' questions with a general poise and mannerliness that indicates their serenity. So in social integration the whole family moves forward, the children the while acquiring both physical co-ordination and social apprehension.

It may be that we shall be criticised for describing many seemingly ordinary events and situations already familiar to our readers from their own experience. To a great degree this is not an account of happenings that are new, but an attempt to draw attention to the importance of such action in human development. It may seem a trifling thing that a family in South London can sit down to supper together after a swim and say "Good evening" to their friends as they pass by, but we believe it is out of incidents as small as this that there is built up social competence —a very complex human facultisation—which educationalists are seeking when, for example, they suggest that the elementary school boy shall have the opportunity of going to a public school. Except for the Centre at Peckham, where in this country can an ordinary man with confidence take his family and day by day among his friends cultivate the art and grace of human fellowship ?

CHAPTER XI

GROWING UP

WE have seen that the children of the Centre have been growing up in families which are themselves embedded in a society constantly increasing in richness and diversity. Here where tastes may be shared in the pursuit of varied activities, each family has been able to create a home-circle with its outer layer of casual contacts and acquaintances and its inner core of friends. Outside this wide circle there is the more general, and for the family alien, amorphous and unpatterned life of the great world, soon to act as a magnet for the child now coming to adolescence. As, step by step, the parents have penetrated the society that surrounds them, the child has little by little shared in their experience, not unlike the way in which as an infant when he began to be weaned from the breast he shared in the food of the family table. So now, again already equipped with familiar experience for a new exploratory journey, he stands on the threshold of adolescence. Eager for exploration and for the taste of more diverse and distant experience, he begins to reach out for new forms of nutriment. So we shall find that the chief sociological characteristic of adolescent development is rapid movement of the boy or girl towards the unexplored field that lies at and beyond the periphery of the home circle.

In our present disintegrated society we cannot of course expect to see a consistent picture of smooth weaning at adolescence. Where the parents in their functional dislocation from society have been egotistically spinning on their own axis, the child may long since have been shot from the centre of the family prematurely to find his own feet in an alien world; or where the parents, long the victims of social starvation, have found no food for their own development, they with binding tentacles will seek to grapple him closer to the family nucleus. In either of these cases, with the onset of puberty we shall expect to find signs of pathological *reaction*. But it is not on such pathological elements of society that we wish to focus attention. It is on the healthy; on those families in which at adolescence there is to be seen a smooth process of weaning leading to a functional re-orientation *of the whole family*.

What have been the broad lines of the child's development up to this point ? As foetus, in close tutelage at the placental bench, working in complete mutuality with the mother, his anatomical structure has been laid down. Then comes birth—the first weaning—which releases him into a wider field of action. Here he continues to be nurtured within the bipolar field of parental experience where the maleness and femaleness of the parents, resolved in the functional unity of parenthood, create a growing 'home'. There, through the long course of infancy and childhood, unmoved by the stir of his own sex, he has been perfecting *the facultisation of his anatomical and physiological endowment—his personalia*.

Now comes adolescence when—as the foetus emerged from the womb into the outer world at birth—he emerges into the unknown world of society. Following a long gestation through infancy and childhood, at adolescence we witness the *birth of the psyche*, with all its potentialities for the establishment of the full functional relationships of adulthood. So youth, now gaining a consciousness of his individuality to the point of spiritual apprehension, begins to exercise his new-found capacity for the conscious direction of his own life in society.

It is unfortunate for a just appraisement of the facts that the development of the individual at this time is linked so vividly with the development of the sex-organs, so that the whole phase has gained an untoward sensual or 'sex' significance. To the biologist the physical unfolding of sex denotes not merely differentiation of certain specific anatomical and physiological features, but the differentiation of the whole individual, with all the attributes of full manhood and womanhood that this implies.

As we look at the process of puberty, therefore, we shall be watching behaviour. Indeed experience has shown us that the most delicate index to the emergence of puberty is to be found, not by the examination of the bodily features, but in the social action of the individual moving in a mixed society. The stir of puberty, which begins at any time from the age of 11 onwards, and is commonly ascribed to sex development in a very restricted sense, is in effect a general development, sociological and psychological as well as physical. It pervades the whole individual giving his or her every action 'bias', and so orientating each anew.

Until this stir of puberty the children's associations, particularly the boys', appear to us to be determined not by their mutual sociability, but by shared interests. Companions will swim or skate together day after day, week after week, and then, if the period of application to a particular activity of one child ceases to coincide with that of his companion, the association breaks down and the two children in their now different activities pick up other companions. Although the 'working partnership' has broken down it is quite usual for there to be no break in friendliness between the two when they chance to meet.

In girls it is not at all clear that an interest in the same activity is the chief reason for their association, for, even as far back as the Nursery, they show a tendency to be concerned with persons and through them to be led to their choice of activities. Watching in the gymnasium or elsewhere in the Centre a mixed group of children congregated round some individual in process of doing something or constructing something, we have almost consistently noted that when that individual moves on to do something else he will leave behind him the majority of the boys, fascinated by the construct or practising the action they have been watching; whereas the majority of girls will follow him (or her) to his next action. It would seem that in the sexes it is their *approach* to action and to things that differs; not the things they do nor their capacity to do them. Certainly the action-pattern of the girls in the Centre is different from and to us more complex than the action-pattern of the boys. These different manifestations, obvious from so early an age, lead us to suppose that there is here an inherent difference needing understanding and demanding experimental study. We are forced to ask ourselves the question—Can it be that our approach to development (Education) in the female child should be not so much through things as through persons? This is a point that can only be studied in circumstances in which there is freedom of action, such as the Centre provides.

But marked as is the difference in the behaviour of male and female children, it is as nothing compared with the sudden change in that of both sexes as the girl or boy bursts into puberty. As a girl develops there is an abrupt assertion of her personality. She will quite suddenly stand apart from her friends and behave in what is for them an entirely unexpected manner. Instead of at once joining them and going with them into the swimming bath as had been her wont, she will perhaps sit alone, swim later or not at all, for she may shed her activities as suddenly and completely she deserts her friends. With her school hat coaxed into some semblance of fashion and her hair arranged in a more grown-up style, she will take up with new companions usually more developed than herself and with them become occupied with their grown-up interests such as clothes, the art of make-up, etc. As she enters the building there now clings about her the flavour of her budding individuality; the passive immanence of her femininity. This is in some cases emphasised by an acute access of shyness or sometimes finds its expression in an almost insolent defiance.[1]

It is not unusual to find that two girl children have for some years been absorbed in the same occupations and in each other's society, unaware of any difference in their stages of development until sudden puberty in one reveals a gap that cannot be bridged. The subsequent abrupt separation is very distressing to the younger[2] child, and may be so to the parents of both children, particularly if they lack understanding of what is happening. The parents of the more mature girl may feel that her behaviour denotes a fickleness in their child. This upsets them and may even seriously disturb a friendship previously existing between the two families.

Sudden and obvious as are these signs of budding adolescence in the female, we are not yet in a position to correlate them with the succession of physiological evidences of puberty; e.g., with the development of the breasts, the onset of menstruation, etc. The exigency of dealing with a constant influx of new member-families has made it impossible as yet to make a detailed study of this subject.

With the onset of puberty in the boy we see a temporary reversal of his previous behaviour with a marked transfer of allegiance from activities to people. He now often leaves his old companions and, finding a group of new associates, hunts in a pack with boys at a similar or slightly more advanced stage of development. He too will leave his previous companions with

[1] Our knowledge of each family usually enables us to relate such over-emphasis of behaviour in this phase to distortion in the previous development of the family *as a whole*.

[2] We do not here refer to age in years, but to biological age—that is to say, stage in maturity.

no trace of regret and may quite suddenly give up an activity that has been his absorbing interest for some months past. The pack now has become his paramount interest, claiming all his attention and loyalty. With the rest of its members he will probably begin to grease his hair, wield a comb concealed in his breast-pocket, and do a bit of surreptitious or even ostentatious smoking.

Such packs are very inco-ordinate and tentative. They seem to be distinguished by no formalised activities and usually become manifest in some sort of rough house in a secluded corner of the building. The pack will suddenly flare into activity on the edge of an adult party; at another time its members will express themselves in cat-calls, or in the sudden seizure of a girl's hat—for small groups of girls are always to be found on the fringes of these packs when they settle down in one place for half an hour or so.

Even if the parents are unaware of the pack activities they are likely to become distressed at the boy's sudden desertion of his favourite sport or other interest, at his smoking and often rather crude behaviour. They feel uncertain about his new companions who seem to them to be leading him astray, and in the homes of both boys and girls at this time there is often uneasiness about what is afoot and what is going to happen.

The observer for his part has difficulty at this stage in discerning the pattern of order underlying the adolescent's action. It does not surprise us however that the action-pattern of the individual should now appear confused and its form undefined, for this is a time when differentiation is immanent.

But undefined though the pattern of order yet is, if the family is considered as a womb with the child a 'foetal' individuality growing within it, then all these signs of personal adornment and of accentuation of personality, as well as the developing aesthetic sense or spiritual apprehension which follow, will be recognised as signs heralding the birth of the psyche, or full and conscious individuality. These signs should tell us of the avidity with which the adolescent is now ready to learn from the adult. The warning too is there. If at this critical time there are only available to him promiscuous contacts and crude evidences of sex, such as in our present disintegrated state of society are only too evident in the films, the books, the dance halls that abound;

if he finds about him no functionally integrated adult society into which he may legitimately penetrate, no invitation to the discretionate use of its appurtenances when his individuality should be emerging in its full expression and virility; then, for want of suitable nutriment, he is condemned to inco-ordinate action, and to precocious and stunted growth.

While the Curator, as he moves about the Centre, is watching the formation of these packs and many other changes occurring in the social excursion and activities of the growing children, the biologist and bio-chemist in the Physiological Department are studying these individuals from the physical point of view. The correlation of physical signs with behaviour in the sequence of adolescent development is one of the subjects scheduled by us for research when time has allowed children with the progress of whose growth we have full knowledge, to come to adolescence.

The material collected by the observers on the social floors and the physiological knowledge gathered in the laboratory and consulting room, flow together in the Family Consultation under the direction of the biologist. Here, with the approach of adolescence, as the whole field of family function is once more surveyed in consultation, the relevant facts and information emerge in their topical prominence, and each member of the family can now make his individual use of them. It is here that parental anxieties and filial impatience can be dissolved; here that the whole family can find the knowledge and understanding to enable them to go forward with appetite and courage in this new phase of development which is affecting them all.

Such a consultation coming, for example, when the eldest boy is at the 'pack' stage, can be of great help to the family. The biologist has seen the pack at work, he is familiar with its actions and its meaning, and has seen the boy's relation to that pack. To him these changes are an indication that the boy is seeking to expand his field of excursion, though commonly they are interpreted as a desire to break from the home—which in the case of the pathological family is probably the correct interpretation. The parents, familiar with all the same circumstances from a different aspect, are able to talk over the situation, and thus gaining knowledge and confidence, reach an understanding which enables them to deal with it constructively without the anxiety born of fear of the unknown. Here is a lively example of the mutual nature of synthesis between member-family and scientific staff of the Centre.

Perhaps at this consultation—though usually at an earlier one—the biologist, armed with knowledge of the developmental state of the child and of the sociological and psychological aspects of the family, will discuss the continuation of schooling for the child and will raise the question of the future career or employ ment suited to him, have the parents themselves not already brought up this subject. Into this discussion the parents enter with interest, resource and goodwill. Many firms employing labour who are in touch with the Centre, use it as an intermediary through which they can come into contact with young workers wishing to train for skilled occupations. These are valuable contacts to us for they enable many of our adolescents to be launched into an industry of their choice for which they are physically and temperamentally suited, and they give a personal link between the boy or girl, the family, the firm and ourselves that may prove useful in dealing with any subsequent eventuality. The Centre staff, armed with a full knowledge both of the individual and of the family in which he grows up, are ideally equipped to fulfil the demands of vocational guidance of the young about to enter Industry and other fields of adult endeavour.

How the child's initial experience of all sorts has come to him as he grows up will inevitably find expression in his adult activities, including his work. Whether he proceeds progressively and in harmony in all that he does, or is hampered by difficulties of his own making that render him temperamental and irresponsible, will to a large measure depend on the family in which he has grown up. Deficiencies due to his nurture may of course be compensated to a degree varying with his own understanding and intelligence. But if the functional organisation of 'family' is the biological mechanism in which the young naturally grow and, through progressive mutuality of synthesis of the whole 'organisation', acquire a physical altruism, which finds expression in social action, then the nature and conditions of the family from which the individual comes will prove to be of importance even in Industry—as modern researches are beginning to indicate.

To return to our observations of the adolescents. Usually in the Centre we see the first pack soon dissolve ; it is only a transitory phase. Following its dissolution, its members either again take up an old interest and carry it on to a new phase of proficiency, or, and this is more usual, take up something new. This they may do with one or two companions only, or they may band themselves together in a new group. Sometimes these groupings consist of boys only, sometimes of boys and girls, and they vary considerably in size and in duration. But this next group is far more co-ordinate and purposeful than the original pack ; its members have a common explicit objective and they are usually prepared to give considerable time and trouble to achieve proficiency. A group of boys in this stage rarely has any other basis for association than devotion to some accomplishment. Certain individuals of the group seem to be always on the fringe, others right in the centre, and there is movement between the centre and the periphery. These groups appear as spontaneous and self-originating zones of activity into which individuals can freely pass in response to the degree of attraction for them of the activity carried on there, or can unostentatiously withdraw to merge again with the general body of society, or perhaps temporarily to consort with some adolescent of the opposite sex. The groups are only exclusive in the sense that any individual attempting to enter for other purposes than that of the activity pursued, is immediately ejected. From the biologist's point of view therefore we should perhaps not speak of them as groups of adolescents, but rather as *foci of activity*, each 'focus' having to some degree a roving capacity. So, for example, we may see a shift from say swimming—which might carry 20 individuals—to an accomplishment like dancing, in which only 15 of these same individuals might be involved ; and so on. Nevertheless the sociological aspect must not be overlooked, for we find that there is at each focus of activity a certain nucleus of individuals all of whom engage in each pursuit shared by the group. Thus it is no surprise to find that at the very centre of the zone there may be—though it is not always so – a leader. This young man whose versatility in skill focuses the other members' efforts, is often a leader unbeknown to himself and unacknowledged by those he leads. He holds his position in virtue of the relevance of his skill to that particular situation. His leadership though real is as ephemeral as the group itself, and he as *spontaneously emergent* in the group as the group is in the society of the Centre.

Throughout this period, as the packs form and dissolve again and in time give place to groups bent on achievement, the important point is that the rapidly successive groupings and activities are the individuals' expression of their advance towards adult prowess and social competence, and of their growing capacity for selection in the social realm. It is of the greatest importance, therefore, that the society in which the adolescent moves should be sufficiently *fluid in its organisation* to allow youth the possibility of making many and varied contacts and, through the formation of successive groupings, to continue its development *at its own tempo.* This can only come about where the young are in contact with a society which includes people of varying degrees of maturity manifesting desirable achievements and accomplishments to master which the adolescent will go to considerable trouble.

It will be seen that all systems which by their very constitution effect a segregation of individuals by sex and age, and exert their influence to consolidate any interest stirred, are by foreshortening the individual's natural excursion, militating *against* his ultimate development. It is unfortunate that nearly all Youth Organisations operate in this manner, regarding continuity of effort and long duration of membership in a segregated society as signs of 'character' and loyalty. To the biologist it would seem that these organisations, in attempting to retain the adolescent in isolation from adult society and thus cut off from the source of stimulus to his next stage in development, are confirming him in immaturity. His continued membership is only too often a sign of arrested development. The biological urge to development however is in fact stronger than any systematisation, for all club organisers bemoan the fact that at the age of 17 or 18 or older, the boys on whom great effort has been expended tend to drift away in search of wider experience. In the presence of a vertical age-grouping of society on the other hand, where the adolescent is surrounded with every stage of maturity and every degree of skill, there is ever present the *natural* incentive to development that satisfies.

During the four and a half years that the Centre has been open, a number of groups such as we have described have come into existence. One of these, studied with particular care, arose as the young boys and girls who as children had been using the

Centre continuously, began to grow up. Dancing, one of the chief cultural, athletic and social activities throughout history, is an important activity for people of all ages in the Centre. The facilities for it are excellent—long uninterrupted spaces, good floor surfaces and skilled dance bands formed by the members themselves. Regularly every Saturday evening there is dancing to a first-class band in the long open hall. Admission is free and open to all members over 14 years of age; the large company is widely representative of the total membership, and the dancing visible to all who enter the building for any purpose. There is ample room for the spectators in the aisles and between the pillars; the general tone is lively and gay, and incidentally the standard attained by the dancers very high. So Saturday evening has always a festive air.

The young boys and girls of 14 years and over gather in small groups on the fringe of this vortex of activity. Occasionally a girl will respond to the invitation of an older woman or girl to dance; the more mature girls will already have found male partners. For a long time the budding adolescent boys, far too shy and incapable to venture on the floor, used to stand by fascinated. After some months of watching, the scene became irresistible and these boys would be found in the darkest corner attempting to dance together when no-one was about. At last they had come to it; they wanted to learn to dance.

But being particularly sensitive to adult society at this time and not liking to appear without competence, in order to learn they had temporarily to withdraw. With that curious efficiency that often characterises what might be described as subjective group action, these adolescents proceeded to deal with their own difficulty. From among the members they found for themselves a dancing enthusiast; he, about four years their senior, was willing, even eager, to teach them. They arranged with the Curator for weekly use of floor space, for the loan of a gramophone and records, and they fixed on the night most convenient for their instructor. In a few months no fewer than fifty young adolescents—boys from 12 to 16 years old, proverbially the shyest and most awkward age—were learning their steps and dancing with girls of their own age with absorbed concentration. By this time their teacher had very cleverly divided up the big corridor-room allotted to them into several smaller sections by the use of chairs, so that the learners could be graded into separate groups between the pillars. He found helpers, girls as well as boys, from among his contemporaries and from among the learners themselves as they became proficient.

Anyone who sees these self-arranged 'classes' for the first time is impressed by the extreme seriousness of the boys as they learn their steps and with the quiet straightforward way in which they invite a girl to be a partner. The girls too respond with gentleness and decorum. Yet in different circumstances these same boys and girls would have giggled and sneered and hung back awkwardly, their whole attitude—a cloak for their unfacultised ineptitude—expressing ridicule and contempt. Nor can one fail to be struck by the fact that these boys and girls are not only learning to dance, but that all unknown to themselves dancing has become for them an instrument of their discretionate social exploration.[1]

This spontaneously created 'dancing class' is then a temporary segregation within a society for a specific purpose. The stimulus to the acquirement of this skill has been the dancing of a mixed community predominantly adult, and we have seen that the means of satisfying the adolescents' desire to achieve this skill, their instructor, was drawn *by them* from within this more mature society. This young man for his part and some of his friends found an opening for a hitherto undisclosed talent for teaching that was most striking. Thus, this situation created in the dancing class fulfilled the needs of the adolescents and gave scope for the capabilities of their teachers, so engaging both in interrelated functional action—a situation which a physiologist might liken to a balance across an interfacial surface, or 'living' membrane.

No better example than this of a *topical* method of education in a mixed society could be cited. We have found that this method applies not only to dancing but to the spread of knowledge and skill of many varieties, especially among those who do not respond to more formal methods of education. Indeed, it must be noted that before the dancing class just described came spontaneously into being, we had ourselves at different times made various efforts, with little or no result, to induce the young adolescents to learn dancing. The important factor in the emergence of any form of activity is the *subjective appetitive*

[1] Photograph 33, p. 63.

stir; where this is successfully aroused, the individual, given the raw material to work with, will himself find the most appropriate and efficient means of accomplishment. This only confirms the principle upon which we are working: namely, that the potentiality is inherent in the organism, and that its spontaneous emergence in action depends upon the cultivation of the (family) environment.

Let us return to the boys and girls equipped with this new ability to dance. We next see them using their new found skill as a means of penetrating further into the society of their seniors. An instance can be given of this invasive process. In addition to Saturday night and party dances, on one other night of the week there is another dance known as the 'Tuppenny Hop', accessible by payment but open to all to watch from the window.[1] To this come young married couples and young men and women who are expert at the most up-to-date dancing. They have assembled a small band of younger members who are masterly in their improvisation of jazz and swing, or, an alternative to the band is available to them in so far as the wireless enthusiasts can broadcast throughout the building, and will through a gramophone turn-table co-operate with the dance group in the absence of a band.[2] Here in this gathering, band and dancers together express their own interpretation of modern dancing. This coterie of skilful exponents is not exclusive but it attracts only those who are also skilful and in tune with young people's life of to-day. Into this slightly older group than themselves the adolescents who have learnt to dance are now able to penetrate. So bit by bit they move deeper into more grown-up society, at each stage their increasing skill in both technical and social accomplishment enabling them to absorb its content, and establish themselves in its milieu. As well as taking part in the 'Tuppenny Hop', they can also play their part now in the more general Saturday evening dancing[3] and on such special occasions as the annual New Year's Eve Party, and in any casual dancing that springs up in the Centre as a result of some boy or girl sitting down at the piano and beginning to play.

The 'Tuppenny Hop' is interesting for a reason not yet touched

[1] Photograph 39, p. 64.

[2] This again illustrates the easy interplay of group with group in the variety of action which is a feature of the life of the Centre.

[3] Photographs 43 and 44, p. 66.

upon. The older people love to watch it and in this way the young bring to them aspects of modern life of which they would otherwise remain in ignorance. It is well known how often people who cannot themselves manage such forms of dancing or who know nothing of its technique, condemn it out of hand. In the Centre however the older members show a lively and enjoyable interest in it, though they are by no means always uncritical. And there is reciprocity even here—in the dance—between the old and the young. Very few of the younger members dance the waltz in the old-fashioned way, but nobody is more enthusiastic in acclamation when the waltz or the valeta is so danced with old-time grace and skill, by some of the older husbands and wives. Here we have a further illustration of the very easy and natural intermixing of the young with the older and more mature elements in society that plays so important a part in the education of the adolescent for living.

There is a great happiness and vigour about these young people as their social capacity grows. They take part in the drama and concert parties, they play badminton and table tennis with friends they made at dancing, in the swimming bath or at the Country Camp;[1] they rally to help at every sort of special occasion and emergency. The remarkable feature of their vigour is that it finds expression not in exclusiveness but in acceptance and participation in the general life of the Centre.[2] Ready at any moment to forge ahead for themselves, they are equally ready to help in any undertaking of the older members. There is mutuality here too, for the activities in which only the young excel—water polo, high and fancy diving, acrobatics, hot jazz—are enjoyed and furthered by the other and older members in their various roles as helpers and spectators. In this way the gathering skill and capacity of the adolescents makes its continuous contribution to the growing vigour and diversity of the society of the Centre as a whole.

But experience leads us to think that the adolescents' association with adults needs to find its expression not only in leisure

[1] Photographs 32, 33, 36, 37, 38, 39 and 40, pp. 63 and 64.

[2] As an instance of this we might cite the invaluable help given by these very adolescents at the outbreak of war. It was they who were untiring in running messages, in making ceaseless journeys between the Centre and the farm, in carrying parcels and heavier loads, in erecting barricades against blast, helping without stint in the manifold tasks incidental to the evacuation of the mothers and their babies to the farm.

but in every activity. Going out to work should play a most important part in the unfolding of adolescence, for association with adults in *responsible* work is in itself an educative factor of primary importance. It is concrete evidence to the adolescent of the growing up of which he is so conscious and of which he so eagerly seeks tangible confirmation.

In this connection we have been very impressed with the difference we have observed in the physique and balance of development of boys who go to work at 14 as compared with those who remain at school until they are 16, 17 or 18. In the former there is an all-round robust functional development, often in spite of adverse industrial conditions, while those who continue at school seem overgrown—rather like an etiolated shoot—as though their development were distorted as a result of the sequestered atmosphere of school.

This was an unexpected observation to us and will, we believe, prove so to the reader. First, it was arresting in view of the almost universal acceptance of the policy of extending school education, with its present age and sex segregation, up to 16 or even 18 years of age ; and second, in view of the acknowledged shortcomings of the modern industrial field in producing what we conceive to be healthy conditions for the adolescent.

But before generalising let us look also at the girls. Our experience so far shows us that the effect of early entry into industry on the female is very different. Whereas with the boys who enter industry young there is a development towards maturity which is to be seen in their physique as well as in their general conduct, in the girls it is marked in many cases by a persistent general immaturity accompanied by a precocious development of femininity. These conclusions can of course only be tentative, but such evidence as we have is striking and conspicuous.

Here we have a subject that needs most careful and penetrating study. Upon the due, sequential and smooth emergence of full facultative maturity of both sexes depends the vitality and so the future of the race, and no educational policy can be surely grounded without knowledge of the biological processes at stake. Neither industry and the circumstances of entry into it, nor the circumstances of education can with impunity be left out of the scope of the health administration of the future.

As they stand, both industry and the schools are inadequate to meet the needs of health for the developing adolescent.

The knowledge, experience and skill acquired in the rich social milieu of the Centre, and inextricably woven into every action, provides a framework for a social education in which the development of masculinity and femininity can go forward to full manhood and womanhood. The desire to become socially co-ordinate seems at first to be the dominant note in the melody of adolescent growth, but once the individuals 'feel at home' in their society—and such an adolescent 'establishment'[1] shows very clearly in their bearing and their actions—they begin to use their newly acquired competence in the instrumentation of their developing masculinity and femininity. This is so unconscious a process that its mechanism can best be appreciated through a picture of the life in which they are immersed as they spend their evening leisure when not otherwise occupied in the Centre.

Let us look at the swimming bath at about 9 o'clock on a summer evening. It is packed with young men and women, some of them already brown after their summer holidays or week-end camping, the girls very smart in their well-cut swim suits, and there is to be seen every degree of good diving and swimming. They are not all young adolescents. As well as three or four engaged couples and some who are 'walking out' there are a dozen married couples, some with their babies asleep in the night nursery below. Nearly everybody knows everybody else. Groups of three or four girls and as many young men are ragging and teasing, and the whole Bath appears one big party.

This 9 o'clock half-hour in the Bath is not, then, just the chance association of those who can manage to swim at that time. For the majority it is the chosen time when either by mutual arrangement or by undisclosed design those who wish to be together go swimming. On most evenings this trio of girls or that bunch of boys or that girl on her own, will stroll up and down and around the bath putting off taking swimming tickets until sure that a certain group or individual has already taken theirs. In the Cafeteria and Long Room there will be some of the younger adolescents who do not swim so late, watching the fun, mostly unaware of, but perhaps not unaffected by the patterns of courting interplay going on in front of them. Some

[1] cf. the 'establishment' of the baby after birth. Chap. IX. p. 171.

of the parents and older members are in the bath too ; but most of these latter are in the Cafeteria, spectators enjoying their brief leisure, or chatting over a glass of beer. It is in this seemingly casual scene where action is unfolding, that we are in fact witnessing the evolution of the idiom of a society and the furtherance of its differentiation.

This nine o'clock half-hour in the swimming bath, then, is an example of a situation we meet so often in the Centre, where we have the interplay of every kind of diversity and degree of maturity among the sexes. In a thousand different ways the adolescents watching or taking part are being familiarised in preparation for the next meal of experience they are to encounter ; in the midst of this society they are, unconsciously perhaps, learning the facultization of sex that is to lead them on to full maturity.

The safeguard against the crudity of action of immaturity that might be feared from such conditions, lies in the example of the facultised expression of the more mature. In the immature responsiveness is still diffuse ; it is most readily elicited by the most obvious stimulus. Just as the infant turns to the brightest light, so in his innocence the young adolescent is first attracted by the more mature, and prefers those older than himself. This 'calf love' is not that which the mature for their part are seeking, so that we get the association with the unattainable from which the gauche unfacultised immature may learn.

We have watched a young girl of 13 who somehow managed to get herself taught diving by the best diver in the Centre, an attractive young man of 21. He happened at that time to be working overtime and was very uncertain of the hour that he could come, but she would sit and wait for him, if necessary for hours. She showed no particular interest in diving at other times, and did not practise on her own. The diving to her was a method of contacting maturity, though had she been a less skilful and promising pupil the young man would probably not have bothered with her. For how many young girls their admiration of a film star expressed in the dog-eared postcard photograph carried in the hand bag is the nearest that they can get to expressing this hero-worship—the natural attempt of immaturity to find its educative medium, the mature. Where in any society maturity is itself moving to selectiveness and discrimination, it constitutes a spontaneous check upon actions of crudity in the immature. So a situation in which mutual interplay is always in progress is in itself an education ; it is the opportunity for the development of discrimination. Perhaps therefore the segregation of adolescents all at the same level of development, when they should be freely learning from every possible degree of maturity, is one of the worst crimes committed against growing boys and girls.

It is, as we have seen, not the stimulus of mere casual contact with members of the opposite sex that is needed to turn the first primitive confused response into a highly discriminative faculty ; but a summation of stimuli ; not just the general effect of the opposite sex but that of continually recurring association with the specific appositeness of individuals of the opposite sex. For it is not contacts alone, nor frequency of contact, but frequency of specific contact that matters—that will lead to smooth and natural development of the sex faculty. It follows that the needs of adolescence can only be met in an established, integrated, mixed community ; not by any segregated group of individuals, not by a shifting populace, nor by the conditions that prevail in the looseness of the great world.

So with their faces now turned outwards from the familiar hearth, we have watched these boys and girls increasingly penetrate beyond the field of family excursion, extending as they do so not only their own field but that of the family with them :—the maturation of the individual and the maturation of the family still proceeding together. Such is the long and mutual process of development that leads to full manhood and to full womanhood :—culmination of the facultisation of sex.

And so in biological sequence we come to the processes of courtship and mating . . . .

CHAPTER XII

COURTSHIP AND MATING

A GREAT deal has been written upon what has been popularly called the "biology" of sex, about the relations of the sexes and about "sex education". In our understanding however, this has been carried no further than the exposition of the anatomy and physiology of the sexes, in its physical and mental expression. What we in this context must look for, is the *functional* expression of sex in the biological economy.

Let us begin with the very simple analogy of the knitter. Nature, like a knitter, weaves her living fabric on two needles—the sexes. The needles work together picking up loops of circumstance from the continuous flow of events, and through their mutual action the materialisation of a specific design or pattern grows. In the knitted fabric this design or pattern is different according to which side we look at—purl or plain. So in Nature's living garment woven by the sexes, there are two aspects, male and female. And these aspects can never be—nor indeed appear—identical, though the 'fabric' of living is one and the same. That is to say, the approach through maleness is different from the approach through femaleness; the *action-pattern* of each is different.

So the sexes do not fit hand-in-glove—one active, the other passive and compliant; neither do they shake right hands in the 'equality' of friendship. Like our bodies, bi-axial in constitution, in their unity they are bi-polar in function—Nature's right and left hands. And that does not mean that the left hand can work the right-handed pattern (or vice versa) which is all that we mean when in common parlance we say a person is "left-handed". True left-handedness would for example produce a script like that of a right-handed person imprinted on the blotting pad—inversed in every sense and only to be read in its mirror image. And this illustration indicates inversion in one dimension only; the complexities of inversion in living in the approach of each sex to every thing, situation and event is something of which we have yet to become conscious.

The only 'equality' between Nature's hands is that, like our own hands, they each have fingers; that is to say, instruments of discriminative sense-contact. But try and put a male glove on a female hand or vice versa and it will only 'fit' if it is turned inside out—eviscerated. The only thing that *will* 'fit' is a stocking-like garment, and that immobilises all the 'fingers', so that they cannot act discretionately! That is the biological price we pay for any attempted 'equality' of the sexes—both hands desensitised and reduced to undifferentiated stumps.

It is true that man is not obliged to follow Nature's lead, for in the individual, sex is a bias that tips the functional balance to one side or to the other, and man can in some measure voluntarily falsify the scales. In the laboratory by endocrine injection he can turn cocks to hens and hens to cocks. At times, disease tips the balance for us, swinging an individual to neuter, or even to the opposite expression of sex. If humanity does not choose to follow the laws of Health, and ignores the bias Nature uses throughout the living world, there are always the laws of pathology as an alternative—and retribution is not always immediate.

In health or wholeness, the bias of sex in man and woman is brought naturally to balance in the unity of *family*. Sex then, is no base 'animal instinct' to be suppressed in man; nor is it mere hunger to be sated. It is the means Nature uses to bring man and woman to adulthood and to the full participation in her life process, for sex is the very *basis* of creative or evolutional energy.

Indeed, so important would the process of mating seem in Nature's economy, that to guide and to inaugurate it, we see her throw the individual back once again under the guidance of *instinct*. Even in homo *sapiens* Nature does not leave mating to chance or wholly to man's choice, for falling in love—herald of the process—is an instinctive and involuntary, or *autonomic* happening. An individual may of volition attempt to stay its consequences; he cannot prevent or anticipate its advent. And, when men and women do fall in love, they are precipitated into a stream of events which in many an instance leads them into paths they had no intention of exploring. This urge to biological completion of the human organism is of immense potency, its strength and delicacy fully comparable to anything yet encountered in the dynamic field. It can move an individual from one end of the earth to the other; can uproot men and women from the binding tentacles of habit and change the tenor of their lives; can release unsuspected potentialities and endow action with immeasurable fortitude.

But in our ignorance we have little understanding, scanty appreciation and inadequate respect for the involuntary or autonomic wisdom that guides this process. We are apt to call it "blind" instinct, and say we have "fallen" in love as though we had tripped over some unobserved object in our pre-determined or 'volitional' path, attributing the incident to the nature of the object rather than to the subjective stir that impels.

In health, that is to say in sanity, instinct and knowledge are not at variance; they are co-operative, voluntary wisdom feeding from the hand of involuntary wisdom (no less than from external sources). So that although falling in love and the courtship which follows are derived from and guided by instinct, we find that these processes are nevertheless not 'blind'. On the contrary, in health they grow clear-eyed and of long and critical vision.

Can we then leave mating to Nature? Man is slow to learn the whole lesson of Science, which is to collaborate with Nature; not to contest her decrees. So he is apt to spurn instinct in favour of his own logic. But the purport of the great process of evolution, of which Man is the thrusting point, is that evolutionary development is proceeding from automatic tropism, through autonomic instinct into 'gnomic' volition. Thus, Man's cue from Nature is to allow instinct to guide the direction, and indicate the content, of his volitional action and to learn to administer his instincts through volition; to regard his instincts as undeveloped *capacities* to be brought through discriminative action to a state of full facultisation.

The biologist cannot ignore Nature's method; hence he must accept and study the processes of falling in love and of courtship as integral to the process of evolution and seek the optimum conditions for their occurrence as instruments of natural selection.

It is with the outlook sketched above that we tentatively approach the subject of mating. It is a different approach from that of the geneticist who already has given much consideration to the possibility of the scientific control of breeding. In dealing with the 'typing' of man's crops and animals, the geneticist has had great success in planned fertilization, hence some technologists in this branch of knowledge would like to see falling in love and courtship ignored as mere emotional illusions, and human mating planned—"scientifically" as it is called.

But it must be borne in mind that up to the present time the geneticist has only been able to manipulate the *physical* endowment of the individual. In crops and animals, the environment of which to a great extent we can limit and control, the physical endowment very largely does determine the results that will follow. In the case of Man however, there is no guarantee at all that he will make use of his physical endowment, or inherited capacities. Most of us are, in fact, very shoddy examples of what can be done with the faculties and characteristics with which inheritance has endowed us. Man is not like the animals in this respect, for his own volition can—and does—deny his potentialities. On the whole, Man has not failed in breeding to physical standards. He has, however, signally and conspicuously failed to cultivate all the faculties with which he has been endowed; failed to nurture what has been so lavishly provided.

It is, then, Man's volition that is the great stumbling block to the application of the science of genetics in Man, and until we have learned to devise a human environment that will ensure the development of our faculties in line with the instinctive guidance of Nature, the geneticist will not be in a position to do much to further the health of the population. His scope must at present be limited to the elimination of inimical hereditary traits encountered in the pathological field.

Everywhere we see Nature using mating to further diversity in the species. With every birth we find her, through the laws of differentiation, arranging for a diversification of pattern, each individual being as unique as is the more static action-pattern of his finger prints. Since inbreeding militates against diversification, the biologist must demand a high degree of variety and free movement in the society within which natural selection is to take place.

In English-speaking society, freedom to fall in love as a preliminary to mating is everywhere accepted. But both parents and society, consciously and unconsciously, use an indirect method of influencing instinctive choice. The method is to limit the excursion of the young within a group radius. This is only too often the operation of taboo derived from fear—and as such is as ignorant as that of any primitive savage—for these groups in present-day society are in the main fortuitous aggregates, largely based on income, privilege and handicap. So there is brought about stratification of society at horizontal levels, and within these contact is largely indiscriminate and promiscuous. Mating, limited to selection within such horizontal groupings, must of necessity result in one form of inbreeding, and it is common knowledge that inbreeding is fraught with peril to racial vitality. The necessary biological conditions for mating that will produce a vital populace can only come from a vertical cultural and social grouping of society.

In civilised man the processes of falling in love and courtship have not so far been the subject of scientific study; indeed, conditions have not hitherto been devised to make such a study practicable. But the time is coming when the scientist will be in a position to write of human courtship as Julian Huxley and others have written of the courtship of birds; for human courtship is as easy to watch studiously as is that of the birds in their haunts, given circumstances in which congregations of adolescent colonies can arise spontaneously in the midst of a society of families, as we have seen them doing in the Centre.

It must not be forgotten, however, that in any group of people in which some 80% or more are subject to physical disorder—i.e. are pathological—the process of falling in love and courtship must be affected by the defects and deficiencies of the participants. We cannot, therefore, as yet expect to see the unfolding of the process in its full significance merely for the looking. Suitable conditions of society and whole, or healthy individuals are essential for this study. Nevertheless though a tramelled upbringing may have restricted opportunity so that the expression of function is but a poor shadow of what it could be; though evidence of what we are looking for must be found in the few rather than in the many; though the individual, acting in a blurred and mangled semblance of the pattern of functional promise contained in his biological potential, will bungle the choice of a mate, or in his non-specific and immature state achieve no more than a group-choice at best; yet the entry of man into his full functional birthright is so magnificent a phase of human development that not all the morbidity and frustration in the world can wholly check its spontaneous manifestations nor mask its splendour. It is veritably the herald of a new birth—the birth of a new adult organism, a 'family'.

Aware of this surging potential capable of development rising in all individuals as they reach adulthood, we as human biologists have chosen this appetitive phase for mating as the chosen time to begin the cultivation of health. It is the new family who must constitute the material of choice for our future observations on health. So it is no accident that we have made possible for such young people an environment rich in diversity of every kind, in knowledge and in action, where a continuity of association can be maintained ; making preparation for their falling in love by cultivating, tilling and composting the social soil, keeping it alive and healthy in order that adolescence may achieve its growth, differentiation and maturation for mating. In this way we hope to increase, not in a season of course, the specificity of mating and thus further the biological process of specific diversification of characteristics.

This is the scientific reason and interpretation of the importance of the social world of the Centre for adolescence, and in this sense the Centre is already becoming a mating and courting ground of salutary significance. Where before joining, a girl or a young man will often have made but two or three chance acquaintances of the opposite sex from which to learn discrimination and from which to choose, after a year's membership the contacts he or she must inevitably have made in the continuity of common action has multiplied a hundredfold or more. And when courtship does come within such a social setting, it need no longer necessarily—and indeed does not any more—find expression in the very formal, relatively inactive and often long-drawn-out process of 'walking out' ; nor is it driven into furtive channels. Instead, it begins and can proceed through action in mutual participation with the chosen partner amidst a society large and diverse enough to allow the courting pair full and varied scope. Moreover, it takes place not under the surveillance of, but nevertheless within easy proximity of the choosers' families, the strong flavour of whose specificity thus can continue to operate as a natural factor in confirmation of the choice each is making. We are all familiar with the wise woman's adage that the ability to introduce the chosen into the old home without embarrassment is one of the good omens of a happy marriage. Perhaps later we shall find that there is a 'physical basis' for this intuition also.

It might well be thought that the maturing adolescents would prefer *not* to conduct their courtships in a club of which their parents were members. We have not found that to be the case. On becoming a wage-earner, the adolescent is free to drop out if he wishes to do so without jeopardising the family membership, or to continue in membership paying his own individual subscription. He thus has a free choice in the matter of his own membership. Experience has shown us that the social milieu of the Centre becomes very desirable at this time to the adolescents of member-families. There is always a lively demand on their part to introduce the newly acquired boy or girl-friend whose family is not or cannot become a member-family. These young people thus sponsored by the sons and daughters of member-families are allowed to become 'Temporary Members"[1] entitled to full social use of the club. Since, however, girl and boy friends have a tendency to come and go, they are not offered periodic health overhaul until they are officially 'engaged' and about to marry. We find it of extreme importance not to cast emphasis, by any provision we make, on the early tentative associations of any couple. The Centre is a testing ground for them at this stage, and it must be as easy, as they grow in discrimination, to end any unsuitable association, as it is to make a happy one.

Among healthy individuals, it would appear that falling in love is usually both a spontaneous and a mutual occurrence. It would seem to represent the individuals' intuitive recognition of the appositeness of complemental diversity in the chosen of the opposite sex. Experience teaches that the healthy adult male's need is for that mutually elective female who can further the maturation of his own maleness, for the needs of the highly specific and mature male are to be met only by the specific, the individual. Mere females—as such—are sexually alien to him. Females in this respect are from the outset even more instinctively and intuitively elective. In health, then, the urge to mate is not an incitement to promiscuous association with the opposite sex ; it is the urge to the *apposite* from among the

[1] These young people, both Temporary Members and children of member-families who are over 16 and have left school, each pay their own individual membership subscriptions—6d. a week, half that of the family subscription —and, like all other adult members, they pay for the activities they enter into.

opposite sex. There seem in fact to be not many 'apposites', for while many men and women can collect cohorts of sex erotics, few men and women are 'fallen in love' with, that is to say, specifically elected for mating, by more than one individual—at any rate at one season.

Neither is falling in love a mere erotic attraction, for it is something that can grip and hold even the most confirmed habitue of sex prostitution,[1] male or female. Moreover, while falling in love is a purely subjective urge, the erotic factor in both the male and the female can be stirred and is influenced by many and various objective circumstances. The medical man is only too well aware of this, for erotic habits such as priapism and masturbation, arising from irritation of various kinds, are met with even in the very young, and are symptoms of a great variety of disorders of the body—worms, urinary and rectal disorders, constipation, etc. These causes, often overlooked, can persist up to and after puberty ; and it is often forgotten that even where they have early been removed, the previous physical habitude can become a psychological fixation. Thus only in the physically healthy can the real measure and significance of eroticism be gauged. In our experience it is a relatively insignificant feature of the total content of maturing adolescence *in health.*

It must be remarked here that were humanity dependent for the continuance of the race on sexual eroticism as is commonly supposed, both the mothering and the fathering of man would rest in very few and very strange hands ! To the biologist, all the prevalent artificial efforts at stimulating eroticism—'sex appeal' and so on—are but one other item in the accumulating evidence of a generalised devitalisation, avirilism and decadence. They are the skilful and deceptive window-dressing advertising an empty and exhausted store—and glamour often a glabrous veneer covering a biologically shoddy constitution.

Whatever the mechanism by which falling in love occurs, it is obvious that more natural and apposite mating will be achieved where the man and woman have developed that high degree of functional action which brings with it discrimination. This

Prostitution is not merely illegal conjugality ; it is the prostitution of *parenthood* with failure to create the new in any sphere, physical, mental, or social. It thus can and does occur even in the married state.

power to discriminate depends upon the previous balanced development of the whole individual, and that, as we have seen in the previous chapter, depends on the mutuality of function of a home set in an adequate social environment, where the social facultisation of the individual can be learnt through familiar nurture. This is concisely summed up in the individual's capability for mutual synthesis with his or her environment; that is to say in the degree of health attained. Thus the health of society and the health of the individual react powerfully upon one another in effecting the instinctive choice of a mate.

In the Centre we watch the healthy adolescent, up to the point of falling in love, revelling in his freedom to move in and about his sociological field, converting every chance he encounters into his opportunity for growing experience. Suddenly he or she becomes attached to one of the opposite sex. From that moment the two revolve around one another; shunning every temptation to act alone. They act as if in a supreme appetitive phase for some new facultisation that is to be achieved. The signs and symptoms of the phase are too well known to need detailed description—of its mechanism, however, we as yet know little. The first necessity is that the physiology of the bodily changes involved in this violent sociological volte-face should be investigated. A study of this subject can however only be made, as we have shown, in environmental conditions in which freedom and diversity exist within an integrated society coming spontaneously under the observation of the scientist. Such a society has to grow and must take time to do so.

In the healthy, between falling in love and mating a relatively long process of courtship intervenes. What is happening during this period? It would be unwise to attribute merely psychological significance to courtship, for, as we already know, most states of so-called psychological excitation are correlated with powerful physical effects. If moods and manners, as is confidently claimed, are inter-co-ordinated with the flow of various endocrine glandular secretions, then presumption is strong that the extraordinary sociological metamorphosis of the healthy adolescent during the courting process is also associated with most powerful physiological changes. If in fear, pain, rage, all-pervading endocrine-secretory changes take place, can we suppose that love leaves us physically, that is to say materially, unchanged?

The practice so universal in bird and animal action of wooing—the billing and cooing between male and female—we already know to be the outward sign of a stir that has its material aspect in the physiology of both partners.[1] We have thus some reason to assume that the human pair, like the animals, are by their close personal and physical association familiarising themselves and physically *sensitising* each other, through this process of courtship. When there has been time to study this subject, perhaps we shall find that this comes about in a manner not unlike that used by the immunologist when, by a series of minute specific doses, he educates the child susceptible to a toxin such as that of the diphtheria germ, to form its own anti-bodies. In courtship, of course, it is not anti-bodies but *pro*-bodies that are being produced.

In the functional courtship of animals, and hence we must postulate in the functional process in the human species, until the final tipping of the endocrine balance there is an almost violent preservation of virginal chastity—presumably to allow for the physiological effects of courtship to mature. So mating is approached by steady and specific translation through a process of familiarisation.

Recent work on the sex hormones attributes immeasurable potencies to the male and female secretions. Moreover evidence is accumulating that the endocrine secretions of the male are related to the endocrinological economy of the female. Perhaps we shall find here also—as tangible evidence of a physical mutuality—that what is excrement in the male is increment in the female. It is interesting in this connection to note that the testicular and ovarian hormones are isomerically related.

We already know that in the male the sperm is continuously available, so that his erotic responses are not periodic, but in continuum, whereas in the female the ova are only seasonally or cyclically available, and her erotic responses cyclically critical, or periodic. Wooing thus would be the penultimate phase in Nature's process of synchronising the coincidence of an inherent a-periodicity with an inherent periodicity ; a yet further example of the mutualising of diversity of function. The result would be that through wooing by the male the crescendo of the endocrine crisis of the female is emphasised, whereas wooing by the female

[1] See chap. II.

dams a settled flow to a crescendo in the male, each thus achieving the peak of coincidence. Certain it is that in the absence of wooing the female tends to passivity in her erotic relations, and in the absence of wooing the male tends to habit and routine —neither of these being an infrequent cause of worry and a visit to the doctor. These are rocks and shoals which often later wreck the mutuality of marriage, for it follows that woman needs wooing not only during courtship, but throughout marriage, and that *she*, not he, is the initiator of conjugation, according to the clock of her cyclic periodicity. We have perhaps some symbolic acknowledgment of this in the custom that it is he who asks her hand in marriage, but she who gives the answer that allows the marriage to go forward.

In falling in love and courtship we see then the indications of a functional process by which the man and woman are actually being physically attuned to one another, in a procession of changes which is to culminate with the physiological metamorphosis of the pair in mating. If this is so, then promiscuous intercourse, whether merely casual, or deliberate as in 'trial-marriage, far from having any use in assisting in making an apposite selection of a mate, is biologically a dangerous procedure liable to confuse the developing specificity of each partner. For we must recall the fact that any 'foreign' or promiscuously introduced substance given in small doses tends to create reactionary allergy, or anaphylaxis. It is thus likely that in promiscuous intercourse we shall find that the *specific* quality of each participant is being blurred, defaced and worn down to a commonality. From the functional aspect every trial that fails merely confirms and perpetuates the error in the capacity to function. Thus, the manliness of adolescent chastity and the womanliness of adolescent virginity may well be no mere ideal of moral philosophy; it may be the expression of Nature's discriminative behaviour in furthering the genius of *homo sapiens*—his individuality and his uniqueness.

It seems then that courtship is the process by which a man and a woman are learning step by step to function mutually, as a unity. Where in health the individuals have grown up in a sphere of altruism, created through the mutual synthesis of family and society, love will go forward as the greatest and most satisfying of many adventures. But where altruism is unknown to the individual, that is to say where an egotist involuntarily and instinctively falls in love, he will be bound to sense an assault upon the defences that his egotism has set up. In such a courtship, resentment may gradually be built up in place of mutuality. So falling in love is not always a pleasure—far from it. Though in health it can be nothing else, in disease it may create reaction, threatening with the fascination of the snake about to strike. It is more than probable, however, that those whom the diseased and disordered 'fall for' in the primary or initial attraction and learn to hate, are specifically those who could have been loved. So that falling in love either in mutuality of synthesis or in reaction and resentment, is dependent on the state of the individual concerned and on that of his society, and not on the object usually designated as the cause. It is rare indeed for the course of even true love to run smooth, for its establishment implies the purging of egotism in the participating individuals, and in the present state of society that is liable to be a major operation.

In the Centre we have succeeded in accumulating both the material and the circumstances which in the course of time should give us the scope necessary to demonstrate the means whereby through courtship the male and female are becoming attuned to each other, and step by step deepening the mutuality out of which unity is born.

As the pair grow in unity, so we shall expect to see their every approach to things, situations and events take on a new aspect, just as we saw two-eyed vision to afford not only a wider but a different and *original* view from that of each eye looking separately. Out of the complemental diversities of sex, 'novelty' will arise as the result of all that they experience as they become increasingly attuned in mutual synthesis. Is there any indication of this in their behaviour ? In health, as courtship ripens, instinct begins to urge them to seek some territory peculiarly theirs, a 'hearth' from which to operate. This instinctive seeking for a 'hearth' we interpret as the quest for a culture-bed for the issue of their unity ; for a place where they can gestate that issue of whatever sort—in its uniqueness. Their 'hearth' is to be for them the material locus of their growing function of 'parenthood'.

After mating it will not merely be the production of children that we shall look for, but 'parenthood' operating to produce a new and specific synthesis *of the whole content of life* that the couple contact in adult competence as a unity. True, the most tangible and discretionate expression of such parenthood will be the production of the child; seal and stamp of their own specificity wrought in very flesh and blood. But this motherhood and fatherhood will be further phases—further differentiations—of the all-pervading function of their parenthood, following—like the dicotyledenous seed leaves—once the parenthood is planted in the soil of home.[1]

With fulfilment of the process of courtship then in the birth of the new family, the old family organisation is superseded; the umbilical cord of the placental association of the old home is broken at marriage. The ripe fruit now drops from the old tree and settles in the earth to form its own nidus or 'hearth' in mutual synthesis with a quick and living environment; and a new 'locus of parenthood' is created. Just as the nidation of the seed in the earth or of the ovum in the womb is essential before growth and differentiation can begin, so the founding of a specific and individual nidus is a primary necessity before the function of parenthood can develop in any young family.

Thus in seeking the *health* of the populace, the *first essential* for the cultivator (biologist) is to see that adequate opportunity shall exist for the easy and natural process of nidation of the young couple in a society which will offer them scope for their subsequent growth and development.

[1] In pathological states, motherhood and fatherhood may be no more than dislocated occurrences happening outside, and divorced from the enveloping process of parenthood; the conception of the child may result from mere eroticism, or from the compensatory action of a non-functioning mating operating merely on the physical plane. Such examples of sterile motherhood and fatherhood—i.e. *unconsummated parenthood*, are disorders of a serious nature, physiological as well as social.

CHAPTER XIII

THE BIRTH OF A FAMILY

Marriage

So, as we watch the moving scene, we see emerging from the more amorphous crowd pairs of young courting couples. Immediately their marriage is decided upon, if one or both are members, then both have the privilege of complete overhaul followed by a joint consultation in which the engaged couple and the man and the woman biologist who have examined them, all take part.

This pre-marital or first 'family consultation' follows the same lines as any other family consultation, and the subsequent talk has a topical bearing upon the stage of development of the pair. As individuals they, like tadpoles, have been in a 'larval' phase from which they are about to emerge. Already step by step through their courtship each has begun to develop, and they are now ready to complete their metamorphosis becoming a complete organism—a 'family'. And this is no intangible 'spiritual union'; it is a biological *reality*, seated deep in the physiology of both man and woman who throughout their courtship have little by little been influencing each other by their proximity and contact, and are now to confirm this procession of changes in mating. It is explained that just as in immunology it is the last massive dose of innoculation that liberates, so in the procession of biological function it is mating that brings liberation of two individuals into their new orientation, and endows them with new potentialities as a 'family'.

Discussion of such questions as the difference in sexual appetite and sensitivity of the male and female partners follows naturally at this consultation. So also does a discussion of pregnancy and its significance in stamping the seal on their unity. Seen in this setting as the means by which they may achieve maturity, having a child is lifted from the plane of easy sentimentality or from the perhaps now more common balancing of expediencies, and comes to appear to them as natural for their own growth as sunlight—of which one can have too much as well as too little. The inference that they quite naturally and quickly draw from this approach is the importance of being in fullest health at this time so that they may enjoy to the full all that lies before them—and especially pregnancy when that comes.

Neither is it merely psychological adjustment that they will look for in their marriage, but a physical and chemical tuning of paired strings that are to sing in harmony. Moving thus towards mutuality of function they may find that as time goes on everything that comes into their ken—even the smallest and most insignificant trifle—will take on a new character; and this will find its expression not only in new vision and new understanding, but in a *new pattern of action*, as the individual point of view of each is resolved into the organismal point of view of the new 'family' they have formed. Everything in the future turns on a knowledge of this supreme and dominant factor in Nature—the prototype of all creation—*parenthood.*

This biological interpretation of the significance of mating is particularly arresting:—to the woman who is apt to feel that in marriage she will have achieved her goal; to the man who is apt to regard his development as parallel to, if not independent of that of his wife. Early a dawning realisation comes to both of them that marriage is no state once achieved to be stereotyped, fixed and held on to in unchanging conformity to a design limbed in the mazy dreams of courtship, nor on the other hand to be calculated up in the bleak terms of economics and expediencies. It is a great adventure, ever changing, deepening, widening, not only with changed circumstances but with the changes that are going on in each of them as their lives unfold in the mutuality of function.

So from the outset this pair are alive to the fact that children are not the only novelty that is likely to emerge from the new function of their coming parenthood; the fruits of their unity may be rich and various and may appear in every aspect of their life, not merely in the intimacy of their personal relationships but in their contacts with their neighbours and in their society in general; maybe in the world of affairs, in business, in politics and perhaps even in budding statesmanship; in the world of thought, of art and science; indeed in every realm they touch according to their capacity—according, in fact, to the richness of the home they build together.

The Home

Let us for a moment leave the young family and turn our consideration to the biological significance we give to this word 'home'. By 'home' it will be seen that we do not mean the four walls of a flat nor a house with garden attached, but *the field of function invoked by 'parenthood'*.

At marriage, a young couple are starting out as a new nucleus of biological potentiality. As this new-born organism begins life in its new environment, like a germinating seed throwing out exploratory root-hairs into the soil and unfolding shoots into the air and sunshine, the pair will seek—not consciously nor intentionally any more than a growing shoot intentionally seeks the light—for material from which they can draw sustenance and grow. As a result of the mutual synthesis which they now effect through continuous incursion into their environment, they will grow out—first in one direction, then in another—with each thrust extending their functional field and rendering the environment bit by bit *familiar*—that is to say, *of the family.* Thus progressively permeating their environment with the action-pattern peculiar to their own expression of 'parenthood', with their own growth their 'home' grows with them.

We call it 'protean', this home that is slowly built up, and Proteus-like it is, for the human family may come to be made up of several highly individualised entities whose excursion is not limited, like that of the plant, to the soil in which they are planted From their lodgment around the domestic hearth these individuals may set out on adventures into every sphere, and the nutriment that they may gather is of a nature varied enough to satisfy every hunger of the questing human spirit. In fact, as the family grows, to such far distant realms of thought and action may they reach, so independent, so dissociated do they seem, the husbands and wives, sons and daughters, that we are content to regard them as self-contained entities, forgetting that, from the functional standpoint, the prospecting tips that each individual sends out are the growing-points of one organism, bringing into the family sphere of influence a new morsel for digestion at its hearth or nucleus—to feed every member of the family.

The 'home' then is no material fabric: no castle walls set against the impact of society to exclude the world. It is the specific zone of functional potency that grows about a live parenthood; a zone at the periphery of which is an active 'interfacial membrane' or 'surface' furthering interchange—from within outwards, and from without inwards—a mutualising membrane between the family and the society in which it lives. This home has its points of progression, like those associated with the tips of the root hairs or the coleoptiles of the shoots. These are the contact-points of absorption of nutriment for the family and they are set between the foreign and the familiar in the environment.

Thus as the parenthood implicit in the unity of the mated pair becomes with the increasing co-ordination of their unity more explicit, or differentiated, so the home they 'grow' about them gathers in extent and content. It is a zone permeated by the *specificity* of the organism ; a zone in which all the members of the family move in familiarity and hence in biological freedom. The home then is as it were a 'body' of potent and specific biological influence wrapped about the family. In health it is cumulative, increasingly specific and individualised as it grows and differentiates through the ever-extending functional excursion of the family.

It may seem far fetched that such dissociated entities as fathers, sons, mothers, daughters, should constitute an organismal entity, and indeed that anything as intangible as a *field of function* should constitute a 'home'. Yet it should not be difficult to grasp. As separate entities, the adult members of a grown family appear to be a mere aggregate of individuals each as self-directing as seems each single bee. But as we have already seen (chapter I), it is the colony of bees which is the organism composed of many discrete working members, or organs—queen, workers, drones—held together in functional organisation. While the hive, like the family dwelling, is the point at which the organism can be observed in its compact form, the 'home' of the bee is delineated by the range and dispersion of the individual bees in their search for nectar and pollen, and can cover many gardens and many fields. We say the home of the salmon is the sea, although its breeding place is in the sandy pool of an upland stream. So we define the 'home' as the zone in which the action-pattern of an organism is inscribed, and within which the biological 'force' or potency of its unity is maintained.

And this, being a zone of biological influence, represents no one-way 'pull'. The family does not carry on this process of ingestion and the gradual extension of the home *at the expense of* a passive environment that is gradually being robbed of its wealth and exhausted of its diversity by predatory exploitation. In function—that is in health—the very reverse is the case. The effect of family action is further to enrich and to diversify the environment, just as the bee by transport of pollen diversifies the variety and enhances the exuberance of the flowers and fruits of the garden—a very clear example of mutuality of synthesis.

So with the human organism in its health or wholeness, the 'home' is to be *measured* not only by the scope of the facultative action of the individual members of the family—their honey and pollen-gathering power, but also by their fertilising and virilising power on society, in fact by the mutuality of synthesis engendered in organism and environment.

Looking through the glass walls of the Centre, as the cytologist may under his microscope watch living cells grow out in a culture medium, it is only a 'section' of the family in action—its leisure activity—that we can see. The function of the human family covers a far wider field. Its home should include not only the excursion of its members to the Centre, but to the factory lathe, to the school, to the grocer's counter, the tennis court, the picture gallery or to the orchestral concert—wherever in fact the individual members of the family are free to move and to cull nutriment. From the biological point of view *all these should be in such relation to the home that each may feed and enrich the whole organism.* In such a way what each does becomes as relevant to the family as are the excursions of the bee to the hive when it delivers the honey. All action not so related to the family-nucleus is biologically dis-'organised', and hence ill-adapted to serve as a nutrient factor for the organismal life. It is only when 'processed' by the family that that which is alien in the environment can be rendered familiar, and hence wholly digestible and not anaphylactic to its growing members. Where all the experience of the members of each family is not thus functionally incorporated into the home, we have a starved and sick society.

As in any pre-marital consultation the talk ranges over the biological possibilities that arise with the birth of the 'family' at marriage, it seems to give the pair foundation for unexpressed hopes and aspirations that might otherwise quickly dissipate in the rude winds of circumstance, or through diffidence—so common in the prevalent foreshortened social environment—fail to find expression. It has been one of our commonest experiences during these consultations to see the young couple look at each other and nod, as if to say—"I told you so"—as though our talk had confirmed some of their own but half expressed intuitive speculations.

Here and there the repercussions of this talk find immediate expression in the nature of the 'home' they set out to build. In this new light, the material things with which they set up house become a means to an end; not the end in themselves. 'Home' is no longer merely the possession of relatively expensive furniture, so big that later on they will have to sell the washstand to get the baby's cot into the bedroom, so highly polished that the parlour suite will form a fetter about the child as it grows, so precious that its maintenance may compete with a more responsible, if less well paid job for the man, and so on. Neither perhaps—as they well might have done—will the pair now delay unduly the coming of their first child lest it should divert their small income from material ends, and become an embarrassing responsibility separating them from each other or interrupting their comfortable and pleasant state of life—for they already grasp that pregnancy is the natural and most potent means to their own maturing.

So here in the pre-marital consultation the family's use of the Centre's 'pool of information' is critically focussed. It enables the pair on the threshold of marriage to set out with some knowledge matched to those of their needs usually described as 'spiritual'. This new family, now sensing that *they* are involved in furthering the process of living, seem no longer to feel themselves just straws tossed in the wind of events, but part of and one with Nature. At this supreme moment of lively apprehension—the time of mating—they move more quickly to a better understanding than at any other time of what they are and how they work. They set out from the beginning knowing that marriage is no estate to be guarded, but a great and new opportunity to fulfil their destiny sustained by the abiding laws of Nature.

It is one of the most heartening experiences to take part in a pre-marital consultation; to see dawning in the comprehension of one or other of the pair the meaning for them of this biological interpretation of their situation; to hear comments which show that they have caught the sense of its import; to note the way in which he takes hold of her arm as they rise to go, leaving us with the unequivocal impression that they have got a grip on something which is going to make just that difference to their whole lives.

'Biological Junctions'

And here for a moment we must digress to consider and to stress the importance of this phase in the life cycle for the furtherance of health. Just as, pollen-fertilised, the ovule of the old plant drops to the ground to form yet a new plant, so likewise at the time of mating we see the dehiscence from the old family organisation and the planting of a new family in new social soil. It is a time at which, *as a result of a natural process*, individuals may shed their skins and leave behind them the furniture and habits of the past.

The life process in organism is not represented by one smooth and steady curve of progressive growth—mere growing bigger from the start. Before growth proceeds there is a period of what is called 'differentiation', when in the undifferentiated, i.e. relatively amorphous primary substance, the lines upon which growth is ultimately to proceed are laid down in conformity with the inherent and specific design of the species. We see such a process typified in the silent conversion of the egg into the high anatomical organisation of the chick within the confines of the egg shell. 'Silent' and obscured from view though this process of re-orientation of living material may be, it is nevertheless crucial to the growth that is to follow. The characteristic of such periods of differentiation in the life cycle is the throwing of the living material into a fluid or more plastic phase—its impressionability, its quick compliance alike to the influence of the intrinsic contained design and to the extrinsic impact of the environment. The phases of differentiation constitute what might be described as 'biological junctions' at which the traveller changes trains and may take another direction from the one in which he has been travelling. The time of puberty has always been intuitively recognised as one of these periods. In this book we have shown that the times of mating and of pregnancy in the family are others.

In studying pregnancy we have indications which suggest that a susceptible diathesis can be changed for the better at such physiological periods. It may in fact be no old wives' tale that we can "grow out of things naturally". It is more than probable that the individual can shed his old constitution during such times of translation—a belief firmly held by the intuitive clinician of the old school. That such a shedding does not occur with any regularity or certainty in no way denies its being able to happen—given suitable circumstances. But we cannot comfortably sit back and hopefully wait for this salutary change to occur, for there is no guarantee that the circumstances of modern society are favourable. Furthermore, just as the influence of the metamorphosis may be for good, so also new vicious constitutional habits may equally and indeed often do supervene at these phases. The point of practical significance is that, as knowledge grows, we should be able to *utilise these physiological phases to enhance health* ; and particularly to reclaim that majority of individuals who in present conditions—as we have seen in an earlier chapter—are in compensative existence.

In this connection it is interesting to note that these periods of differentiation, the biological junctions at which the direction and habitus may change, are just those periods which are neglected by medical, educational and social workers alike. True, sporadic educational efforts have been made to touch the adolescent, but up to now they have intensified his isolation from the body of society ; a procedure which we have seen is inimical to the natural development of his biological potentiality at this period. Courtship and mating and the conceptional periods of parenthood escape social and educational influences of every sort. Maternity and Child Welfare we have ; both come too late. And as for mating and marriage, there remains only the marriage ceremony, an empty, almost mocking symbol which once stood for the serious care and concern our ancestors devoted to this crucial period of the birth of a new human family.

Growing into a man or a woman—or later into a family—is of course a natural process, as growing into a flower or fruit is natural. But the plant cultivator pays especial attention to the soil *before* planting the young plant, and *before* its flowering. So likewise the social soil in which the human organism grows, demands the *maximum care and attention that we can give at these natural biological periods of differentiation.*

This problem of cultivation is a biologist's not a pathologist's problem. To the biologist the use of the 'biological junctions' constitutes the *primary* feature of his technique in the cultivation of Health.

The Early Days of Marriage

Let us return to the Centre and, using it as an observatory, watch the young family after their marriage. Though the pair are now united, this new organism is as yet wholly undetermined in its functional organisation. We are not surprised therefore to find that in the early days of marriage, for a time they take no conspicuous part in the life of the Centre and from the social point of view are largely lost sight of. This seeming withdrawal reminds us of nothing so much as the state of the fertilised cell in the brief period of apparent quiescence after conjugation before the almost frenzied re-arrangement of its chromatic particles as it begins to develop. The pair have in fact for the time being gone into what might be called a *centripetal phase* of function.

This preliminary withdrawal from society may be one of the reasons why young married couples have not hitherto come within the ken of any social administration whatsoever; why they tend to disappear from 'clubs', and why they have so conspicuously escaped all educational influence at this most impressionable of periods. It must also be noted that it is a stage at which all too often development becomes arrested, and from which it may never proceed, except, perhaps, through the accident of an unwanted pregnancy which ruptures their social encystment, leaving a fragmented 'membrane' of contact ill-adapted to functioning in mutuality with their environment.

Although they may seem to shun society at this stage, we find nevertheless that they are eager for further understanding of the process that is enveloping them, and are anxious to make use of the information that the Centre holds for them. And here it must not be forgotten that such facts as the Centre has to offer are becoming available to these new-born families *in circumstances which make it possible for them to be acted upon, according to and in measure of each family's capabilities for action.* Very unlike the situation of the average young urban family, when they are ready to emerge into social life and activity, they find themselves already embedded in a society rich in possibilities, by no means fully probed yet already in some measure familiar to each of them. This offers them a jumping-off ground when they do burst into the growing phase of their new-found unity. The Centre with its abundance of unorganised 'raw material' is there for them as something to be rendered contextual, to be defined and made specific in the home that they will grow.

Here then in the Centre is a situation in which *the 'practice' of Health in its positive cultural aspect becomes for the first time possible :—family culture on a rational basis, beginning with the enrichment of the soil before marriage.*

We have given a rough sketch of the progress of the family going forward in conditions in which there are many more chances of experience, or 'food' for physical, mental and social digestion, than can be found elsewhere in modern urban civilisation. We must not suppose, however, that the Centre is a stage which has all the 'properties' that any young family could use. Far from it. The properties available are only such as we, in our experimental essay, know how to provide ; such as our very restricted finances allow us to provide ; and such as the circumstances of modern life beyond our control permit.

In the four years of the Centre's existence these young married couples have been all too few for statistical study and the time all too short to see any but the first fruits of their subsequent development. There has been no time for the children of member-families growing up in the Centre surroundings, to come to mating and pass into maturity. All that has been demonstrated so far is the alacrity with which those families who come early enough use the knowledge that is available, and the certain courage and assurance with which they have begun to build up a mutual family life and home. In contrast to what we find in those joining the Centre four, six, eight years after marriage, this in itself must be counted positive evidence of the value of a cultural approach as a means of enhancing Health.

CHAPTER XIV

SOCIAL POVERTY

In the Centre the smooth lines of functional action, or health, begin to stand out as the high-lights of a relief wrought of the living medium of the family membership. But the very prominence of this 'action pattern' of the family, illumined as it is for us by a hypothesis of biological function, only serves to accentuate and to deepen the shadows of the background from which the relief rises. Thus, by the very sharpness of the contrast, we gain not only a knowledge of function, but also a clearer understanding of the nature of the negation of function—its pathological anatomy and, more important still, its aetiology or origin.

Everyone knows of this dark background of habituated and reactionary existence. Never before, as in the Centre, has it been tracked to its lair, curled in the embryonic or new-born nucleus of the home; nor have the stages of its growth been displayed in their sequence under the very eye of the observer :—the child conceived of parents in whom no functional unity has been established, parents in whose bodies before conception deficiency and disorder lie cloaked from knowledge; the child, epitome of potentiality, carried with diminishing vigour as it grows in the womb, to be born into nestless nakedness and, in the starved travesty of a home, reared in uncultured bondage; at last, ill-nurtured, to escape into the unfamiliar welter of a society disintegrated as the family from which he issued, there to breed his like again.

Every week families in one stage or other of such dissolution join the Centre, many crossing its threshold only to quit—as quickly as they come—an atmosphere too strong, too vigorous for their ineptitude, their wilting strength or short-lived efforts. The post-marriage history of any such family—should it manage to stay at all—is written clearly for us in the physique of each one of the individuals that go to make it up and also in the general conduct of the family as a disrupted unit; as well as being told in the story that grows out of the family consultation and their subsequent use of the Centre.

Social Starvation.

There was the young married woman, just about Flo's age, whose husband's work made it impossible for him ever to come to the Centre with her, but for whose benefit he eagerly welcomed their membership. She came several afternoons a week for two or three weeks. Alternating between fear and shyness, she sat watching the bath and the constant stream of activity in and about the cafeteria. She declined politely—and one felt with a sincere reluctance—any invitation to enter into anything that was going on. "No, I couldn't". Contorted with shyness, hesitantly she could utter but one short halting sentence in acknowledgment of any attempt to draw her into conversation or into any casual tea time group. Still she sat until one day her eyes filling with tears, she rushed from the building never to be seen again. In the loneliness and starvation of her unfacultised social encystment, she was unable to stand the strain that the presence and sight of social action put upon her. Drawn to seek company, this young wife was driven by contact with it to the limits of physical flight. And this is no unique case; hers was but an extreme example of a young family planted and left to grow in an impoverished and barren social soil.

On any afternoon there are to be seen in the Centre many tea time groups of three, six, ten women, talking together in easy friendliness. It is still as astonishing to us as it is to the visitor who sees them for the first time that, if questioned, nine out of ten of these young women will answer—"No. I hadn't a friend before I joined the Centre".[1]

Out of Nature's ample endowment, the young urban family builds through no fault of its own, not a rich protean body—a *home* that grows out from the nucleus of parenthood, but a poor hovel of sleeping and eating, breeding and clothing. For all too often the family holds no converse with the outside world; its functional scope is restricted to its own hearth and there is little to sustain and to feed its members but what happens within the four walls of the house. Compelled thus by circumstance endogenously to consume its own products, the exploratory tentacles of the family are withdrawn, and, shrunken around its nucleus, there forms a hard resistant crust of suspicion and defence. In the poverty of present-day social life, even those with healthy

[1] Photograph 15, p. 57.

appetites are forced in upon themselves, the family being driven to sustain itself egotistically—instead of altruistically in mutual synthesis with its environment. Thus is the soil prepared for the reception of the child!

As the couple fail to make contact with acquaintances and friends, and as they fail to find a natural social field of experience, they entrench themselves in the subterfuges and substitutes not of intimacy but of contiguity. So there are most surely laid down the foundations of all their subsequent problems and difficulties with themselves individually, with each other and with their children.

For some families, even for some quite young families as we have seen, the chances for development opened to them by the Centre come too late. But not for all, nor even for the majority. The power of development and the power of recoverability vary greatly from family to family, usually, though not invariably, decreasing as age advances. Let us describe one or two families as they were at the time they joined, and trace the course of their development as we have been able to watch it.

Mr. and Mrs. X had been married seven years and had two children aged four and two. The wife was fat, flabby, constipated, dressed in slovenly clothes, was suspicious, diffident and negative. The husband was overweight, irritable with his wife in private, in public alternating between over-confidence and shyness. The elder child was undernourished, anaemic (lack of night sleep and worms), bit her nails, was a wriggler, a bed-wetter, lisped, was furtive, and when among the other children conspicuous for her inability to share in any of their activities. The baby was fat, of uniform pallor, with a rubberoid slightly sweating skin; was overclothed, inactive, listless; though two years old was unable to walk steadily and unable to speak except in words ill-pronounced; was carried everywhere, all attempts at adventure being restrained by its parents who pacified it alternately with sweets, dummy or biscuits.

Before marriage the husband had been a keen boxer, used to go to "the Club" three nights a week, had several men friends, and was a swimmer—all of which had been foregone since marriage. The wife had been a shop-assistant and had never helped in the dignities of the house before her marriage. Now incapable of achievement—still less of adventure—in house keeping, cooking or in any social direction, she had no sense of home-making and no friends. She had been ill during her first pregnancy, injured at the confinement, and had "never been right" since. She was incapable of managing the children, the elder one, aged 4, being already beyond her.

What had been their life together since marriage? The husband, a vandriver in steady work, got up at 6 a.m., took his wife tea and made his own breakfast before going to work. The wife got up at 9 or 9.30 and gave the children their breakfast about 10 o'clock. She cooked a desultory dinner of one course; walked up and down the shopping street three or four afternoons a week with the baby strapped into the pram, the elder child dragging beside her; went to the park once a week, speaking to no one and coming home early because the elder child was fretful; knew no-one except for a passing acquaintance with the grocer's wife round the corner, and her in-laws, with whom she was on the defensive and did not find herself in sympathy. The husband came home to a hotted-up meal, stood for half an hour or more beside the cot of the elder child—who refused to go to bed or to sleep if he were not there. So at 9.30, or later, he went off to the billiard hall for a little recreation or to the pub till closing time, leaving his wife to see that the children came to no harm. Or, in another family where the husband was too kind to leave his wife on her own, *both* stayed at home every night and had done so ever since the first child arrived. He taciturn, sat with his feet on the mantleshelf; she brooded on his silence in resentment. There was nowhere to go together in the evenings except "the pictures" and they could not leave the children in the house alone. So, knowing no-one who would "mind them" they never went anywhere. Night by night their long drawn out silences were cloaked in the blare of a thirty guinea loud-speaker.

So all this husband had done since marriage was to work. And for what? Both he and his wife were disillusioned; life to them had become stagnant, if not sour. There are no sensitive tentacles nor lively growing points from such a home, feeling and thrusting their way bit by bit into the nearby environment; tasting, testing new experience and affording a continuous flow of nutriment for embodiment in that home. No experience comes to this would-be unity and all its members alike are the victims of functional starvation.

And these are not pictures of families standing out vividly by reason of their rarity. Varying but slightly in detail, this story of early married life is repeated with *monotonous regularity.* It is however the final chapters only that are generally known, and they are to be found written down in full in the notes of doctors, psychopathologists, almoners, social workers, child guidance specialists, magistrates, probation officers, solicitors—all of whom are continually encountering every form of chronic and acute disease and disorder in the home; disharmony, neurasthenia, inebriety, 'suburban neurosis', causes for divorce, parental neglect or incompetence, the difficult child, the young offender; suspicion, retreat and anti-social behaviour of all sorts. The reader will know only too well how common all these are—and in all classes of society.

Experience in the Centre has made it very clear to us that the origin of these troubles lies, primarily, in the nature and circumstances of mating and early marriage. It is as though, as in the picture we have given earlier of the infant not weaned on to a fuller diet in due season,[1] the young family equally starved of suitable nutriment in its appetitive phases for adventure, develops not only the physical pathology of inanition, but a psycho-pathology as directly contingent upon deprivation as is the puling and contrariness of the partially starved infant. Underfed too long, a refractory period ensues for the family, and the disturbance spreads over the whole field of action displaying evidences of physical, mental and social pathology alike.

Here is the story of Mr. and Mrs. X., the family cited above, once they became members of the Centre. The father joined the boxing club, boxed and within four months was found instructing a group of boys. At the first family consultation the mother was referred to a gynaecologist for her disorder to be adjusted. Thereafter she went to the "Keep Fit" class, learnt to swim, meeting various people and becoming friendly in the course of her struggles in the water. Both husband and wife lost weight, moved about more briskly, looked healthier; eventually both played badminton together with their friends, and built up a social circle of their own, coming out perhaps two evenings a week. The husband became Secretary of the Badminton Club; his wife's dressing improved enormously—she even attempted, with the

[1] pages 177, 178.

help of a friend, to make herself a summer frock, a thing she had never thought of doing before. Much care was given at the Centre to the eradication of the elder child's infestation with worms. She gained weight, mixed with other children, got over her wriggling and bed-wetting and on being put to bed went to sleep at once without her father's attention. The baby was brought to the Nursery five afternoons a week, and within a month was walking firmly and sitting up at table with the other children. At 2½ it was going to the gym and to the Babies' Bath. It acquired a colour, its flesh became firm, its mentality active, its personality almost winning.

This is a concise record of the change that came about in this family, but it gives no picture of the gradual and often painful stages by which they climbed, now with hesitation and now with over-eagerness, out of the incarceration and ineptitude to which all alike had succumbed. Satisfactory as was this progress, it was clear to us that, having encountered these opportunities late in their marriage, the parents' actions were not more than compensative measures, and that it was too late to affect the basic constitutional make-up of the elder child and possibly of both children. Had these opportunities been available at the time of courtship and early marriage, how different would have been the home that family would have grown about them!

It must be emphasised in this connection that these changes towards more healthy action which occurred as a result of membership of the Centre were *unassociated with any change in the wage level or other economic factor operating upon the family.*

Let us take another family of better education:—a fine upstanding man of 28 and his wife three years his junior, a good looking girl of great charm and dignity. They have been married three years and have a small girl of eighteen months. The husband is in one of the departments of the Civil Service, where he has the possibility of rising. She, formerly an expert accountant, is a capable woman with a considerable accomplishment as a singer. Both have distinct social gifts; they have a well-appointed flat and a politely mannered child, though too strictly kept and parentbound. An occasional friend comes to their flat, they make an occasional return visit, but the wife has no adequate outlet for her gifts as a singer nor as a hostess, nor any chance of helping her husband to take the place in the world of which she knows him capable. Both have a diffidence not to be explained on first encounter, and the man lacks confidence to seek the promotion he merits.

They are very fond of one another and they are making every 'sacrifice' for the child and its education—an unconscious compensation for their intuitive sense of their own lack of fulfilment. He is at times irritable with her, and what is worse to her, with the child. This makes her critical of him. Their intimate marriage relations are difficult and infrequent because distasteful to her. They feel greatly the responsibility of bringing up children, and, ambitious for the career of their only one, are eschewing further children. It soon becomes apparent that the wife is not far off a nervous breakdown; tears are very near her eyes at the beginning of their first Family Consultation.

On joining the Centre this family move about shyly and discriminately for the first few months. They dance together occasionally but only gradually make new friends. They join the Concert Party; they learn to fence. During this time their marriage relationships begin to smooth out and deepen owing to a better understanding of their difficulties gained at the Family Consultation. At the next Christmas Party the wife sings, with great success. Their life is opening out; they come to know more people. He grows in confidence. By the end of the second year he decides to seek promotion, works for his examination and passes—which he might well have done three years earlier . . . The next autumn by mutual design they decide to have another child. They now make use of the 'medical' department both before and during the pregnancy to make sure that they are as fit as they can be, and that the wife's reserves are being maintained. All goes well. They have been able to make use of the knowledge and the other chances the Centre holds for them. With the husband's now increasing scope in his job they move into a still wider social circle. Meanwhile quite unconsciously the wife has become an educative force with the other mothers among whom she moves with an easy grace and a fund of good sense. At the end of the third periodic health overhaul they look back on their life: "How little we dreamt three years ago that we should ever be doing all the things we have done in these last two years!" The Centre came in time for this family. Life has acquired a verve and a significance for them; they are no longer afraid of it.

The great potentiality of this family had failed to find expression for want of any natural flow in society. Which way a family will move, whether towards a fuller development, or downhill into apathy, social starvation and disease, is determined by the social soil in which it grows. The Centre has demonstrated that this soil is capable of cultivation.

Ebb of the Instinct for Nurture in Womanhood.

In the periodic health overhaul we have found a higher proportion of untreated disease, disorder and deficiency in women, particularly between the ages of 20 and 45, than in men.[1] So also in the social life of the Centre, we find the deficiencies of a relatively inflexible and inadequate social environment afflicting women more than men. In the unhappy state of modern urban life, the man has his legitimate work and though it often fails to give him the scope he needs and the opportunities he could take, and though it leaves him in most cases with little or no leisure to share with his family, his interests are usually just sufficient to protect him from the stagnation which afflicts the rest of the family. He is comparatively well off. With the woman it is different; marriage is apt to lead her into conditions in which social stagnation and inaction are almost unavoidable, even have circumstances allowed her to develop her potentialities before that date—which is far from usual.

When we piece together information gathered from the periodic overhauls, family consultations, and from observation of the actions of the growing girls and young women in the Centre, we gain a sorry picture of the functional development of the average young artisan woman coming to marriage. She is of an age to have been born at the time when a whole concatenation of events occurred to bring about far-reaching changes in the domestic life of her parents and their home. It was a time when there was rapid increase in the wages of her parents; when the six-roomed cottage was replaced by the three-roomed flat; when cheap goods began to flood the market so that the making at home of the man's shirts and socks and the children's clothes became unnecessary; when lack of accommodation made the

[1] See chart I (*frontispiece*)

taking-in of washing impossible and when the commercial laundry and the bag-wash met the new need and ability to put the family washing out; when the quick gas-stove running to the tick of a slot-machine replaced the continuously burning kitchen range, and at the same time cooked and tinned foods requiring the minimum of skill in preparation, became universally available. These changes, all coming together as they largely did, suddenly released her mother from the burden of her earlier domesticity. With less work and more money, but with *no further education and without increase in suitable outlets* enabling her to use her new-found leisure, this mother, previously attuned to ceaseless toil suddenly, found herself in a void. With what result? The focus of her attention and interest turned in upon her children, on whom she now began to wait hand and foot. Moreover—and perhaps more significant than all other changes—about the same time the large family was coming to be replaced by the small one, or even by the only child.

So it is not at all uncommon to find a girl—throughout childhood the sole focus of her mother's attention—who as a schoolgirl of 11 or 12 is wakened by her mother with an early cup of tea! She has never been allowed—far less expected—to do a hand's turn in the work of the house, and has never even mended her own clothes nor cleaned her own shoes. She goes from primary school to secondary school or commercial college and becomes a "lady clerk". Her wages enable her to dress smartly; she goes about in the car or pillion riding with her men friends; she is able to go to many dances and to pay for a yearly holiday away from home. Sailing through her young life under the auxiliary power of maternal solicitude she has lived without any need whatsoever for knowledge or experience of domesticity. Its arts are a closed book to her; often distasteful, dreary and difficult.

What then happens when she comes to marriage? Not only does she bring no "bottom drawer" on the preparation of which she and her mother have been dwelling since the time of her adolescence, but she brings no *dowry of skill*. Leaving behind many of the pleasures of the old life, she comes to a new, and what functionally should be a fuller life, without any particular gift or aptitude for it. No apprenticeship served about her mother's house, no indwelling upon and preparation of trousseau have measure by measure stirred what may well be the endocrinological responses of this girl: and the education she has received in its place being modelled upon that of her brother—man's education—has not fulfilled this natural need intuitively recognised by the wise women of the past. So, in her life precast in the present mould, no *appetitive phase* for this knowledge has ever stirred.

After "walking out" for several years, or spending her leisure and her wages with her boy in the public dance hall—where there are aggregated a crowd of dissociated entities, socially unrelated one to another—and the cinema, she leaves her old home and marries. The pair furnish their two-and-a-half room flat 'en suite'; often on the hire-purchase system, so that there is noth ng left for the young wife to do to complete its equipment. They live either on pre-cooked or tinned foods, or suffer the girl's unaided experiments in cooking. These may be amusing at first, but soon become discouraging and the attempt quickly slips into an easy and monotonous lack of inventiveness. By the time the baby comes—all too often an unwanted accident—we find her still ignorant of how to make even a milk pudding or a soup for its weaning.

The prevailing ignorance of the young mother in domestic matters is becoming generally recognised. The remedy usually prescribed is courses in cookery and housewifery. In this connection we must give our experience in the Centre. During the fourth year, two courses of cookery were arranged, one for those accustomed to use electricity, the other for those using gas in their own homes. These classes, the arrangements for which were made in co-operation with the mothers in the Centre, were held at a time convenient to themselves and particularly at a time when young mothers were in the building with their infant children. The classes were well-attended—but by whom? By the middle-aged mothers who already were tolerably good cooks and wished to improve their skill. Of the young mothers, less than a handful came—in spite of the fact that they were in the building and at leisure.

How different was the response to the suggestion that the young mothers, at first with help and subsequently on their own responsibility, should undertake the preparation of the Nursery teas for *their own children*. This includes the preparation of vegetables and salads, and the cooking of such simple dishes as purees, junkets, jellies, stewed fruit, etc., with many of which they were not previously familiar. Needs prompted by marriage and by the coming of the baby can lead the young woman, not through lectures or precepts from the expert, but through action with those in the beginning unskilled like herself.

Nor, indeed, has she any experience of babies, for in her home there were but two children ; not the large family from which her mother came and where there was always a baby with the handling of which the older children early became familiar. No amount of technical lectures or training in Welfare Centres can replace this intimate experience of youth. So the sense of 'nesting' and of home-making remains rudimentary. The experience is unfamiliar, the appetite for it is not there—and in its absence the young wife is often blithely unaware of any inadequacy. Perhaps worst of all, she comes to marriage with a sense of having *attained* her goal. The implication this brings with it is that there is nothing for her to do but to continue in this state sustained by a loving husband.

In this picture pieced together from the various sources from which information comes to us in the Centre, we have a red light, warning indicator pointing to the trend of events in the life of the modern girl which leave her instinct for wifehood and motherhood unfacultised.

Let us look more closely at her life after marriage. Her husband is out all day. Her housework takes her an hour or two at most ; her shopping another hour. After that she has *nothing* to do till 6 o'clock when she must think of preparing the evening meal. She has lost sight of her school friends, and now lost the associates of her work and has no means of making new ones. So, not wishing to become intimate with the other people who live in the same house, she has no friends. All too soon time begins to hang heavy on her hands. Perhaps if she is vigorous she decides to go on working for a while—the extra money will be welcome for she has been used to having and spending her own wages. That puts off the coming of the baby. Or she may, as did one young wife, travel ten miles daily to spend the day with her mother, so keeping happy and occupied. But this young woman had little or no desire to alter these circumstances of life, and the marriage was not even consummated when two years later she and her husband joined the Centre.

Joining the Centre turned the tide of development for this pair. As a result of the talk at the Family Consultation they were anxious to adjust their conjugal relationship. Eighteen months later they had their first baby. What a change was effected by that pregnancy! The girl gained in physical vigour and assurance, she established herself in her own motherhood, and the focus of her attention moved from her mother towards her husband in their mutual parenthood. Here were two people who had been living in compensative existence, complaining of nothing when they joined the Centre—not even of their unconsummated marriage—who, finding a source of information and an environment favourable to development, stepped out into an active functioning life.

The Declining Birth Rate.

As year followed year and the members themselves came to recognise the changes that were occurring in their friends and acquaintances through successful pregnancy and parenthood, we came to be increasingly involved in the use of modern substitution therapy, both endocrine and nutritional, to induce pregnancies in those previously infertile. In the course of our work we have found evidence of what we believe to be a high percentage of infertility, of non-consummation of marriage and of rarity of connection, as well as of the deliberate avoidance of child-bearing by birth-control methods. From the observations made, we feel bound to link this incidence of non-consummation of marriage and of relative infertility, with the question of devitalisation. The nature of devitalisation has been discussed by us elsewhere.[1] The type described under the name "hypotonia" found in a much higher proportion of women of all ages—even the very young—than in men, is a definite physical condition with definite physical signs such as we have described. These have their accompaniment in the mental and social fields of action (or rather of inaction) of the individual. In these people what is usually presumed to be the pressing urgency of sexual appetite appears to remain unstirred to a surprising extent. In our first survey[2] this group, imperfectly understood, was included amongst those exhibiting sexual continence. In our

[1] "Biologists in Search of Material," p. 63.
[2] "The Case for Action," page 69 et seq.

later work we have been in a position to make a further analysis of this subject. In the Centre, in the society of which there has already begun to occur a gathering desire for children, these young people are beginning to disclose to us the origin and nature of their infertility—infertility of which often they themselves are only now becoming aware. It would seem as if there were in them a positive lowering of all appetites, a natural fasting—i.e., a raising of the threshold of the response to stimulus. Just as this type is work-shy because of a relative incapacity, so they are sex-shy and for the same reason. It is as though these people were insistently and continuously faced with some compulsion not to exceed ten miles an hour, and the disillusionment they suffer in crawling along at this pace in a machine designed to do 100 miles an hour is a cumulative depressant.

From this analysis of involuntary infertility, we have come to see that much voluntary infertility—i.e. avoidance of child bearing—is not necessarily, indeed not usually, to be attributed to selfishness or vice, nor even to rationalisation, but to a barely conscious recognition—by the woman in particular—that in her devitalisation, pregnancy may be but one step further towards her undoing. Perhaps the very attitude towards pregnancy—one anticipatory of sickness and danger—which the medical profession has assumed, is no more than a reflection of the type of patient so consistently encountered by the obstetrician! The devitalised individual tends naturally to move further and further away from living towards the protective level of personal survival, that is to say from the biological (actional) level of function to the pathological (reactional) level of encysted existence. These devitalised individuals are existing continuously on a minimum current account; they possess no reserves, no deposit account, no guarantee at all that if they set out on the shortest functional journey they will reach their destination, still less return home with achievement. Intuitively, then, they turn aside from any environment in which something new may confront them. They encyst themselves in their houses, using them as mediaeval keeps with the draw-bridges up because of their foreboding of their relative incapacity to contact any change in their self-limited environment. This *safety-first* attitude to life is an important factor operating on the birth-rate.

Death is Cumulative in the Family.

And what does this attitude to life portend for any child who does happen to be born into such a home ? After the wife has recovered from the shock of conception and her possible early attempt to get rid of it,[1] the baby—which she did not want—now comes to be welcomed, probably as the physical result of endocrine changes at last stirred in her by the pregnancy. So the months preceding the birth of the child become on the whole happy ones. Now, too, she has something of importance to occupy her in the preparation of her flat, of the baby's clothes and of her own clothes. But when the baby does arrive her unpractised hands are tremulous. Its every cry is an anxiety ; its every movement her concern and duty to anticipate and to still. At last her days are full. But, as the baby grows and the time approaches for it to venture on its own, she unconsciously now begins to hold it back. Secretly she is pleased if it shuns a stranger, for that shows its attachment *to her* ; and she broods on the time when it will no longer rely on her for everything, intuitively sensing that she is incapable of moving with it and will fall back into the position of isolation she was in before it came. So in devitalisation and dread of the emptiness of her life she clings to the child, and in the habituation of its debility and thraldom it clings to her. Neither develops ; both become distorted, and a spoilt child is reared—"nervous" in temperament like its mother.

So as this child grows, instead of meeting through the continuous excursion of its parents a rich, varied and lively environment, it is caught in a situation in which there is one unchanging constant—its mother. Not an inanimate constant either, but one whose intent is always *towards the child.* We all know what happens to any entity subjected to one insistent influence. Trees that grow near the coast in the path of the prevailing wind

[1] Early attempts to terminate pregnancy are in our experience considerably less frequent than was the case ten to thirteen years ago, when the first Pioneer Health Centre opened. The reason for this is abundantly clear. Whereas in 1926-29 it was but few who were conversant with the possibility of control of conception, and fewer still who successfully practised it, our later experience in 1935-39 disclosed that the use of contraceptives is now practically universal, and the degree of success attained apparently high. We are, however, not in a position to attribute this apparent success wholly to the contraceptives used, in the face of the indications we have given which lead us to suspect an increase in infertility, both relative and absolute.

are misshapen because, in their otherwise varying environment, there is *one* constant and unchanging factor. So this child never stands on its own feet four square to the winds of heaven, but is directionally set by its dependence upon its mother.

For the mother the danger of fixation is even greater than for the child. In her growth up to marriage she may not have been as a uni-directional tree, but grown straight and in balance. But, if as a result of her marriage all other winds die down—if she loses what friends she had and is without the chance of making new ones, if she is socially isolated, cut off from all work, sports and interests, and devitalised to boot, then when there comes the prevailing wind of motherhood, the only one to blow and to gather as the years go by, all her subsequent growth must be onesided. She will become misshapen, and contact with her will deflect the growth of the child at each of its successive appetitive phases. Mother and child, then, each to the other becomes a constant in the environment and as a result each grows one-sided. None of the buds to windward ever develop and those that grow in the lee of the wind grow *in* its path and in the shadow formed by the tree away from the sun. Here then is a perfect uni-directional tropism. It is the picture of the aetiology of mother-boundness—a condition that cannot be remedied merely by dealing with the child, for its roots lie deep in the family.

In the majority of children of nursery age whose families have joined the Centre, we have seen signs of a greater or lesser degree of this mother-boundness. The only ones who seem free from its stigmata are the few who come from large families where the mother is always busy, and some of those who come from 'social problem' families—and these latter suffer from other ills.

Let us watch these 'mother-bound' families in their use of the Centre's afternoon nursery. We have described the Nursery as yielding a fresh field of experience for the child, while at the same time the activities of the Centre lead the mother, and through her the father too, into a wider field of social activity. So the Nursery plays its dual part in relation to the progressive *mutual* weaning of them all. But that is not its aspect for all the children or all the mothers who join, for where weaning from the skirt has been delayed or resisted by the parent, the Nursery does not—indeed could not—appear as any opportunity but rather as a threat of separation.

Yet to the women to whom the swimming bath or badminton, the congenial circle and a quiet cup of tea, still make their appeal, the Nursery is an inevitability that she accepts not only for the good of the child but in order that she herself may swim or join in any other activity. For days, sometimes even weeks, both mother and child will suffer tortures. The mother lingers with the child when she takes it to the Nursery, inventing excuses to delay leaving while she vainly tries to stop it crying; she only half enjoys her swim and hurries from the changing room to peep into the Nursery or hang over the balcony rails to listen for her baby's cry. Gladly she accepts offers of other mothers to go and see if it is allright, and as early as possible rushes down to fetch it, giving it the welcome of two separated souls who have both been through some awful purgatory.

But if in spite of this she persists, then after a week or two—the time commonly varies from a week to a month—she is rewarded by discovering that she has enjoyed her swim, and only suddenly has remembered that it is time to collect her child; finds that she has forgotten her anxiety, and that the child too now leaves her eagerly to go to the Nursery, meets her smiling, and has become happy and active in her absence. This is the woman who is able to use the chances the Centre offers, as a remedy at first, and once the ill is remedied, as a straightforward opportunity. Thus she lifts *herself* and so her family with her out of the category of compensative existence into the category of health. Without the Centre how soon would she have dragged herself and them all into the category of frank disease?

But this is not the only response. There is the woman who, when faced with a swimming bath feels her own disability so strongly that she cannot bring herself to the point of going in; whose response to every new chance is yet one more withdrawal and refusal. Honestly persuaded of her child's delicate constitution and 'sensitive' nature, unable to bear the anxiety of its not eating its tea, worried about the harm that constant crying might do it, after a few days she will stop bringing it, and so save herself from the painful operation now necessary to take the place of the natural process of weaning. This will probably bring the family membership to an end.

This mother-child relationship is but one aspect of what is a *family* disorder. The reader can for himself appraise its significance for the husband. Other symptoms begin to appear in the syndrome. The young wife's attention, at one time given to her husband, is now concentrated upon the child. At first the father is happy to linger with his wife in her contentment, but as the child grows, he expects to stride out again with her at his side. But not so the wife. She is now tied to her baby, indeed physically as well as psychologically tied, for the baby cannot be left alone in the evenings while its parents go out even for a short walk. A separation of interest begins. It is a small cleft, but one deep in the nucleus of the family—a rift in the unity of the pair.

As the months go by, either the husband stays at home, lapsing into morose silence, or he goes further afield in innocent compensation for the interests and companionship he has lost. But he is changed by marriage. He cannot now rejoin his old bachelor friends, so he falls quite naturally into the company of other disconsolate husbands—birds of the same moult as he. The wife misinterprets this and resents what she feels to be his indifference to her and to their child. This resentment grows in proportion to her concentration on the child, till at last she extrudes him—its father—from the fragmenting home and 'protects' the child from his influence. What a pitiable misunderstanding and how little—at the right time—might have availed to change the course of that family's history. Later we meet the same sort of family living in domestic propinquity in which the wife accepts the material obligations of her domesticity—and even of her wifehood—but has grown poles apart in feeling, intent and action. The sole link between them now is the only child pulled both ways by the severing influence of each parent—subconsciously, if not openly, at war in the barren spaces that should have become a live and functioning home. The mutuality in which a *home* can *grow* has long since disappeared.

This is the picture of many families when in middle age they join the Centre. Owing to the restricted circumstances of their early life they have been unable to draw sustenance for their growth, or to grow from their schismatic nucleus any developing home.

Often the husband has become so confirmed in his compensative habits or negativity that his wife has the greatest difficulty in persuading him to join at all, though to do so would confer on her and on their child some real and immediate benefit that he would be the last to wish to deny them ; or even though he may recognise that membership would help them in dealing with the problems common to them both, presented by their children already out of hand. Or again, the woman in her encystment and unfacultised state may have contracted an obstinancy or unwillingness for change so powerful that the greatest tact and patience has to be exercised by all concerned to persuade her to yield to her child's importunity.

The parents of the school child often present us with a picture of this sort. There are those parents who never come to the Centre except for their overhaul ; those families where the husband never goes out with his wife or the children. A child of such a family once said "I *hate* our garden, my father never comes to the Centre". In other cases the mother never comes, pleading that she is too busy, or pleading nothing at all ; just keeping away, using her membership for its usefulness in occupying the children and keeping them off the streets. On rarer occasions we have met such a mother who later in life has taken up some outside activity with the zeal characteristic of compensatory action. In either of these cases where each parent is acting individually, there is no understanding of the fact that the exploration of the social field by the child they both still cherish, needs for its digestion the specific and familial nurture derived from their mutuality of action. Habit now binds them ; it is too late.

Yet for all it is not too late. We have seen even middle-aged families who were drawn together by the social life opened to them by the Centre.

When they joined Mr. and Mrs. S. had been married for 16 years. Mrs. S. was quiet, competent, but self-abnegatory and withdrawn. Her husband was a man of some considerable native intelligence long unfed in the narrow confines of their limited home. But his energy was not to be denied and in the billiard hall he had sought the society which the terms of life in his district had not elsewhere yielded. There he had fallen in with companions who made use of his unemployed energy and surplus of unused skill, turning them to doubtful purposes. Debt, drink, betting, sharp practices had brought him—and thus the family too—all too near the arm of the law. So the children had grown in the furtive knowledge of something amiss, and the wife, devoted yet unarmed against an intangible situation she was impotent to change, struggled, wearing herself to the bone.

This family joined the Centre. The man, soon recognised as a skilled craftsman, quickly gained the respect that was his due in a society where his gifts could find legitimate exercise. He was quite ridiculously proud to see his small daughter of four and a half sweep round the arena on skates—the youngest child to achieve ease and competence in this form of balance. He was deeply interested to find that something could be done for his eldest son whose disability they had accepted as inevitable. He gave his skill freely and without stint to the construction of the electrical installation of the theatre and the illumination of the Centre for festive occasions, and so on. He never failed to take part in and was a valuable contributor to discussions on the management and progress of the Centre.

After a year all the old habits had dropped away. When it came to their third periodic health overhaul the wife left the family consultation with tears that revealed her deep recognition of what the Centre had meant for them. This family, through the chances the Centre held for utilisation of the man's capacity, has refound its unity and established its equilibrium in society. The husband now has a field for the exercise of his native capabilities, his wife a sphere in which she can operate as an effective focus of their social life and their children a home through which they can be nurtured to the limits of the excursion possible to them as a family.

Social Remedy.

Age does not necessarily deter people from making use of a wider environment. On joining, many older people make magnificent efforts to overcome their disuse-atrophy, physical, mental and social. But, fascinating though it is to us and to them to see their ineptitude in some measure overcome, at this stage in their lives such action must, for most of them, be in the realm of remedy rather than the true development of health. It may be thought that we are being unduly pessimistic or even harsh, but no-one can doubt the truth of this observation who notes the covert interest of the spectators and the embarrassment of her grown-up children, as the grandmother of over 60, less than four months after first entering the bath and learning to swim, makes for the diving board and, amidst a breath-taking hush, dives from the topmost 7 metre board. Such a delayed repair of an adolescent omission is not the balanced action of health. That much of the activity of the adult society is in this realm there can be no doubt, because so often the first reaction of the adult is not, like that of the child, an instant response, but first a drawing back followed by a set determination requiring an effort out of all proportion to the significance of the achievement.

Consciousness of disability is the great stumbling block to the majority. In that phrase is summed up the effect of years of lack of excursion, lack of exercise of potential capabilities and limitation of effort. Faced with a simple situation like a cafeteria full of friendly talk, a swimming bath and swimmers, a dance floor and dancers, their first reaction is so often "Oh, I couldn't do that". Action has been inhibited and now help is needed; the encouragement of friends, the aid of a husband, instruction in a quiet place hidden from general view, a crutch to overcome lost faith in their capacity. They need some stepping stone so that the minute scrap of viability represented by their hesitant desire shall flourish in a soil *not too strong*.

Outlook on Teaching by the Professional.

It is for these people that what little formal teaching there is in the Centre is provided. Apart from one part-time swimming instructor largely engaged in teaching the older women, and a very successful "Keep Fit" class for women,[1] there is now in the Centre no professional instruction. This does not mean that there is no skill that is up to professional standards, nor that the adept do not teach. Indeed teaching and learning goes on busily everywhere, but by the neighbour who can and wants to do it. So teaching as an art of the enthusiast—not of the professional—has begun to flourish all over the building: in dancing, fencing, badminton, in diving and swimming, in dramatic and concert party work, in music, in wireless, in dressmaking, cooking and so on. Indeed professionalism in the Centre has proved to be not only unnecessary but actually inimical as a

[1] Photographs 41 and 42, p. 65.

means of encouraging the development of skill in the *ordinary man and woman* hitherto without skill.

The consciousness of disability that we have described permeates the adult's approach to all kinds of activities—even the simplest—to reading, to writing, to dress, to all forms of social excursion. To such people nothing inhibits their own attempt at action more than the demonstration of a high degree of skill. So it comes to be true that *in the present unfacultised state of society the exhibition of high proficiency is a bar to adventure by the majority*. Specialism will always defeat both the educator and itself if it is allowed to flourish unchecked, extending its sway beyond the field of the enthusiast and the already partially competent. Any activity exclusively organised on a basis of skill will fail to attract the interest and sustain the endeavour of the man-in-the-street, for the first essential to gain these ends is to sweep away his unconscious self-distrust.

Potentiality Wasted.

Working in the conditions created by the Centre, it has become abundantly clear to us that what in these older people lies unexpressed for want of opportunity in the past, still has some degree of viability. Though learning to swim or to dance at the age of 60 might be the tardy repair of an omission of youth, the energy to do the repair is an expression of a residue of health. This power and wish to achieve is poignantly moving. That these people should need to exercise such effort and suffer so much to achieve a co-ordination proper to youth is a tragedy indeed.

These families, the blasted oaks, the pollarded willows, and the stumps in the forest of the society of families we see in the Centre, are all too prevalent. But it could hardly be otherwise when we consider what their lives have been. Though the shoots and branches will burst through the bark and clinging corky layers of the old trunks, such sporadic outbursts, proof enough of what the development might have been, cannot now contribute to the tree-like contour of the family.

Amongst our members there are, of course, the few who have become so confirmed in their sedentary habits that no fresh growth is possible. They try to make use of the club solely as a place of amusement, and grumble if there is not constantly some ready-made entertainment to take their fancy; members who show a total inability to make use of the varied chances for action that the Centre offers, or who when they join pick up some previously established interest—generally their only one—and for two, three or four years show no change in that interest and never move outside their chosen circle of cronies. We have, for instance, members who play whist two nights a week, three when we have an extra game, and in the afternoon also when we have an afternoon drive; who cry out for it on other afternoons and evenings and who go to other clubs where they can get it at times when it is not on in the Centre. These families, though conspicuous in the company, are in a small minority.

Youth.

What of the children of these middle-aged families? The first thing we notice is that how the weaning of adolescence is accomplished affords clear evidence of whether the growth of the home has been healthy, balanced, 'whole' or whether pathological and distorted.

In cases where no functioning zone of mutuality ever seems to have formed within the family, at an early age the children appear as though extruded from the family, being left to fend for themselves prematurely in an alien and unfamiliarised world. After the first two years, members of the staff who might not know the circumstances of any particular family, came to be able to spot such children as they moved about the building with what might be described as a waif-like mien. They seem in some way without any centre of gravity in the community. They are the products of atrophic parenthood. In the same type of family at a later stage, where the adolescent has persuaded his more or less indifferent parents to join, but where apart from the necessary visits for periodic health overhauls they never come near the building, the adolescents are usually 'off the rails' and characterised by bad behaviour.

Where, on the other hand, we see the children cling to their parents, where they are diffident and afraid; or where on the contrary they are over-eager to break from the home and headstrong to dash into an unknown world; also where the parents fight against their going, regret their departure, or attempt to impede it, so that the weaning of adolescence becomes a battle royal, we have the measure of a parentbound home.

Often the father, but more particularly the mother, clings to the children as the only link with the outside world. Social life is so constricted and so moribund that by the time their children reach adolescence, the parents have no field of excursion. There is no useful work for them to move on to when their children are fledged. In five years those working in the Centre have become keenly aware of the problem of the competent mother left without occupation when her children grow up. There is here a store of wisdom and ripe experience which we feel should be turned to good account in the community.

From reference to the comprehensive list of activities in the Centre,[1] it will be seen that there is as yet little that makes a particular appeal to the older man or woman. But in spite of this, they do come, for the age distribution of use of the Centre's amenities, is as wide as that of the actual membership, and the Centre has a fine robust middle-aged life. Every new activity carried on by these adults is to them a social adventure :—their contributions to the organisation of the various activities, the ideas they bring and the work they do to make the Centre better known in the district, their new found ability as swimmers or badminton players, their dramatic or other interests—all carry them forward to a wider life. Not only does the spice of life thus enter once again into the mutual relationships of many middle-aged couples, but it enters into their relationships with their children. And so it is that we often see them beginning to relax their too tight grip on an adolescent or older son or daughter.

But how often the absence of any natural outlet causes the parents to become parasitic upon their children! We have a number of families of this type ; families where the boy or girl comes only with the parents. It was a pitiful sight to see the girl of 17 who, if she did not enter the building with her mother, was inevitably followed by her within half an hour ; to see her charming and coquettish advances watched and approved by the mother, who, kindness itself—the iron hand in the velvet glove—never interfered but who had such control that there was never any danger of anything happening that needed interference. One had the impression that never for a single moment was her daughter outside her consciousness and it was evident that she was never beyond her physical ken or psychological

[1] See Appendix I.

power. Or, we may cite the boy of the same age who, held at home by homework enforced upon him by his father's ambitions, went out rarely in the evenings and then only with his parents; who left their side to play billiards—his father's chosen game—returning to his parents obediently when finished and whose independent movement and initiative was reduced to a minimum.

These children's lives are being lived *for them* by their immature and *insufficiently occupied* parents, whose energies should normally be flowing in other directions. In later stages the action of some of these children is significant. We see them at last react against their parents' domination—perhaps they either never come if there is any chance that they may meet their parents in the building, or when they meet them studiously and sometimes insolently refuse to acknowledge them. But whether they react or not, it is probable that the majority are scarred for life, their own undevelopment being carried over into their own subsequent marriages.

Youth and Leadership.

The two extremes, the neglected and the overlain, are few in number—and equally distressing to meet; but those in some degree affected are all too many. We know what the world at large is doing to *correct* these pathological adolescents; it is shouted from the housetops at every Educational Conference and at every Youth Committee. First it segregates them, then isolates them from the opposite sex, tends always to isolate them from all older and younger members of society—and then cries out for "leadership". But there is no short way to eradication of disorders that have their origin deep in the family circumstances. It is the social environment that is defective; the social environment that needs cultivation—not "leadership". It is forgotten that the natural leaders of the young are to be found in society, where every skilled man, every amateur athlete, every happily married couple, become automatically and —most important—unconsciously their leaders.

But herd these adolescents together and incarcerate them in age groups cut off from the natural incentives and inherent discipline of a mixed and more mature society, then a situation is created in which *both the stimulus to and the control of action must be provided by authority*; by masters, not leaders. That measures are taken to palliate and adjust the condition of the adolescent thus only deepens the disaster and leads to the conclusion that the present trend of the social services for these young people is actually furthering the devitalisation of society —unwittingly, of course. The stream of social life having been allowed to become blocked and stagnant, to further isolate the adolescent is to produce localised whirlpools in the remaining fairway into which finally even *the vital* are sucked.

In making plans for reconstruction—too often without reference to natural law—it is hoped by means of remedy to save a situation the origin of which lies in long-tolerated circumstances that have bred a disintegrated society. From this there can proceed neither sustenance nor natural stimulus for the development of the family. So the educationalist and the social worker —like the doctor with a new found drug—finding a remedy effective in certain clamant cases, hasten in optimism or often a kind of quackery, to apply it to the populace at large. The inevitable long-term result of this procedure can be no other than a devitalised and 'drug'-sustained society. This danger into which we are so prone to fall is, in fact, similar to that of the accepted practice of feeding the populace with artificially prepared vitamins and other substitute 'foods', to make up for deficiencies that arise from agricultural procedures which, denying natural law, lead to a continuous depletion of the *vital* qualities of the food produced.

The alternative—in each case—is *to plan for health*, whether health of the soil or health of the society in which the human family is to grow; reserving *remedy* strictly for those who fall by the way. The demands of sickness are so vociferous and cumulative that they distract our attention from the truth that even a modicum of health is a *leaven* which, if liberated to act freely in society, has power to 'vitalise' the whole lump.

Bad Beginnings.

What chance then has the young man or woman growing up in the segregation of a disintegrated society to find his or her appropriate mate? The way to increase the possibility for appropriate mating, as we have seen, is to *widen* the excursion and multiply the contacts of those who are inevitably going to mate. For whatever the conditions, mate they will, falling in love being almost the only vestige left of primitive instinct in the adult human being. The poverty and starvation of the conditions in which social contact among young people of both sexes can be made is hardly appreciated, and rarely referred to. It is, however, in our opinion, one of the root causes of unsatisfactory mating and of the subsequent inco-ordination of parenthood. So restricted are the conditions for meeting and association, that we are forced to witness on every hand vicious forms of social inbreeding—and without thought or heed to the genetic effects we go on with the process of sex, age and wage-group segregation of the urban and suburban populace and call it social and housing reform. We have here a glaring example of the specialist who in 'curing' social disorder, is unwittingly sterilising society.

And so the young are driven to backstairs, cinema gloom, and sub rosa courtship. This is fraught with follies and disaster because it accentuates and unduly exaggerates erotic attraction, so clouding and narrowing the choice and inducing reactionary return to purely compulsive mating ; whereas choice, to be choice at all, must arise out of the interplay of discretionate action on the part of adult individuals. But discretionate action in adulthood is the result of the progressive facultisation of the individual, who stage by stage as infant, child and adolescent has found circumstances in which he can pass naturally from one appetitive phase to the next in the exercise of appropriate action. This can only arise as the result of a nurtural process in which he moves from the intimate and familiar in ever-widening excursion in the midst of an integrated society—such a society of families, for example, as we have seen beginning to form in the Centre.

The Place of Psychology in the Peckham Experiment.

One last word. The Centre has often been accused of neglecting Psychology. In our opinion, however, there is as yet no psychology ; only a knowledge of psycho-*pathology*. Indeed, how can there be a scientific study of psychology until an experimental field for the *study of the healthy* has been established ? At present, knowledge of the physiology of the psychological mechanism lags far behind its pathology which fills libraries, for all knowledge so far available on this subject is derived from the disordered, and its theories and technique are devoted to remedy. Remedy is important for the sick, but the sick are not the primary concern of the Centre. Once however a medium is found in which we can watch the action-pattern of function in the family unit, we are presented with an experimental field in which we can begin to look for a psychology based upon health rather than pathology. That such a field simultaneously gives us the opportunity of observing non-function and of tracing the departure from function as the individual parts company from health, is true, as this chapter demonstrates. In this way we have come to see displayed before us the aetiology of the 'devils' to the more advanced stages of which psycho-pathology has long been paying attention. In the Centre we see clearly that the disorders that both we and the psycho-pathologist know so well, are no more than the tail-end of a long train of events following upon unfulfilment of the functional development of the family—in due season and sequence. With the figures for *physical* disease and disorder before us, it is no surprise that we should find corresponding disorder and absence of function in the psychological and social realm. Moreover a disintegrated society must inevitably deprive the family of a live environment, its legitimate source of nutriment for development.

It must be recalled that for this experiment we chose what we thought likely to prove the least disintegrated society to be found in the London area. Its disintegration is not due to poverty, unemployment, abnormal sickness, the congregation of one class or of one wage level. It was, so far as we could judge when choosing a specimen of the populace to work with, the most healthy and vigorous. This is all-important in view of the picture we have had to give of the background of families against which the first sketch of the action-pattern of health has made its appearance.

We cannot reiterate too often what our experience has taught us. This picture with which the Peckham Experiment has presented us of *social* pathology with its accompanying psycho-pathological signs arises out of primary disorders just as apparently trivial, just as effectively compensated and therefore cloaked, and of just as long duration, as the physical disorders we find in the bodies of those who, in the aggregate, go to make up the families we have been considering. Neither can we repeat too often that such minor disorders would either not occur at all or would fail to develop into the grave socio-pathological conditions we have had to describe, in circumstances that made a functioning life possible for the young family.

It is not wages that are lacking ; nor leaders ; nor capacity ; certainly not goodwill ; but quite simple—and one would suppose ordinary—personal, family and social opportunities for knowledge and for action that should be the birthright of all ; space for spontaneous exercise of young bodies, a local forum for sociability of young families, *current* opportunity for picking up knowledge as the family goes along, and the seemingly trivial—"Oh, I couldn't do that", "I couldn't go there" that spell out so long and truly pathetic a story, and such deep-seated psycho-pathological frustration, need no longer be heard in the realm.

Health is more, not less, infectious and contagious than sickness—given appropriate circumstances in society for contact. By that we mean that Nature has its *own* laws of transmission for health which are no less inexorable than are those of transmission in the realm of pathology. But health, whether through intuitive means or through knowledge, can only come through *conformity with these laws.*

CHAPTER XV

SOCIAL SUFFICIENCY

THE Centre is an instrument devised to throw up health from a sample of the populace, as a separator throws up cream from milk. This book has in the main dealt with the appearance of health as it is beginning to come to the surface. The large residue of 'not health' has been dealt with only in the last chapter and in the first two sections of chapter VI.

The day to day work of the Staff however is necessarily determined by the nature of the total content of the Centre membership. It is not our concern to treat disorder, but that disorder should be removed is a necessary preliminary to the cultivation of health. So it comes about that we are in continuous contact with every type of therapeutic agency willing to deal with members recommended by us for treatment. In this way the Centre has been the means, not only of referring many of its members to the appropriate source of treatment for their physical disorders *at the earliest moment they are detectable by a trained diagnostician*, but also of putting them in touch with public and other auxiliary services with the existence of and means of approach to which they were often ignorant: e.g. the Ivory Cross, the Hospital almoner, the Charity Organisation Society, etc. In this way many families have been saved much waste of time and needless anxiety as well as much protracted ill health.

But there has also fallen upon the Staff the responsibility for dealing with a wide borderland of conditions never hitherto systematically encountered by any therapeutic agency, and for which consequently *no diagnostic or therapeutic provision at present exists*. Often, too, where provision of this kind does exist, the circumstances in which it is procurable are such as needlessly to interrupt the daily life and work of the individual. This is something to which he will not submit—and which would be detrimental to his *health* were he to do so. The individual, as we have shown elsewhere,[1] from the youngest to the oldest, seeks to maintain himself to the last minute in his social situation, however poor that be. Even at the cost of considerable suffering, men and women will not "give in" until they must.

[1] *Biologists in Search of Material*, p. 78 et. seq.

The Health Centre is the first and only organisation that openly conspires with the individual so to sustain himself in his society, while at the same time taking deliberate measures for the elimination of his disorders—even the most trivial. This of course at once wins the individual's confidence and co-operation, but in the present circumstances in which the therapeutic agencies are not yet organised for the reception of disorder of those not manifesting the *symptoms*[1] of physical or social disability, it also throws onto the Staff much work for the reception of which in the future the therapeutic departments will be organised.

There are many instances in which, from the point of view of the cultivation of health, it appears important to eliminate some minor disorder or abnormal tendency, but where to the therapist the disorder seems to be so trivial as to have no importance, or on the other hand to be so intangible that it is beyond his scope with the instruments and circumstances at present at his disposal. As time goes forward these difficulties will solve themselves, for were the therapeutic agencies to be consistently supplied with the early case, which at present does not reach them, they would undoubtedly devise an organisation to suit its needs.

A second factor of importance in determining the balance of work of the Centre's staff is the fact that the longer the duration of the Centre's influence in the cultivation of the health of its member-families, the greater will be the body of health among them. So year by year there must inevitably be a consistent shift in the consultative work of the Staff, away from the interruptions caused by the necessity of eliminating disorder, towards the elaboration of technique for the cultivation of health.

Of whom does the Staff who carry out this work consist? For the 1,200 or more families with whom we have been in contact in the past four and a half years, there have been working five biologists, three or four with medical training and experience —senior and junior doctors; one or two with a science training, acting as curators of the instruments of health; one bio-chemist with assistant, a certified Sister-midwife, a receptionist, two nursery supervisors, a cafeteria manager, a general floor manager, a secretary, and students of biology attached as occasion

[1] In medical terminology a "symptom" is what the patient complains of; a "sign" that which the doctor finds wrong.

demanded to various members of the Staff. All except students are whole time workers present daily from 2—10.30 p.m.,.Sundays excluded.

Five days a week, Saturday evenings included, the periodic health overhauls go forward from 2 to 10 p.m., by appointment at the convenience of the member-families. Daily all the equipment of the Centre is available for all its members. It has been a strenuous four and a half years for the Staff confronted with entirely new material and having to approach that material with a new outlook. Sustaining the routine of examination alone demanded organisation and the evolution of a technique of approach, which was exacting. To those medically trained, the discipline required to concentrate on health in the presence of such a wealth of new material of clinical interest was rigorous. It often implied a unique clinical opportunity personally to be foregone. But perhaps the most difficult task of all has been to refrain from following the many lines of special research, clinical as well as biological, that unsought have opened up, constituting an ever-present temptation which in the beginning had to be resolutely resisted. Our first necessity was to work out the general lines of procedure essential to a health organisation, and to learn the type of approach suited to health practice. Only now are we beginning to be ready to inaugurate within the Centre special researches, the material for which is ripening with such abundant promise in many directions: in medicine, in welfare of every sort, eugenics, psychology, education of the family and of the child, economics of the family; and above all in the nature of health and its cultivation.

A review of any week's work by the Staff of the Centre will disclose the range of material covered in their routine. It will be found that they have handled work which in the ordinary circumstances of present-day administration is carried out—usually at a later period in the history of the individual—by a wide range of public and charitable services. The following is a list of activities both cultural and corrective covered by the Centre for its member-families.

Welfare and Educational	*Therapeutic*
Ante-natal Clinic	Marriage Advisory Bureau
Post-natal Clinic	Mothers' Clinic (diagnosis only in the Centre, with personal reference to source of therapy)
Birth Control Clinic	Child Guidance Clinic
Infant Welfare Clinic	School Care Committee work
Care of the Toddler	Poor Man's Lawyer
Nursery School	Hospital Almoning
Immunisation Centre	Hospital Follow-up, including all forms of after-care for *all members discharged from medical care*
Medical Inspections of the Schoolchild	Rehabilitation Clinic
Vocational Guidance	
Sex instruction of adolescents	
Girls' and Boys' Clubs	
Youth Centres	
Sports' Clubs and Recreational Clubs of all sorts	
"Keep Fit" and Gymnastic Classes	
Adult Cultural Education : Music, Debates, Drama, any classes or lectures desired by the members	
Citizens' Advice Bureau	
Holiday Organisations	
Outings and Expeditions of every sort	
The Public House	
The Billiard Hall	
The Dance Hall	
Social Gatherings	

For the carrying out of all these activities the Centre organisation has been found to be peculiarly well adapted for the following reasons :—

1. The Centre being a *family* organisation, and its varied activities being carried on in such a way that they convert all situations into material for the self-education of its members,

both the family life and the social environment in which the family moves is being continuously enriched. Working not in an admonitory, and not in a palliative fashion, the Centre is thus a local cultural (i.e. developmental) factor in family life.

2. Since the Centre is the families' own Club—of which its Staff are the servants—the members suffer no patronage, no loss of prestige, no sense of inferiority and no undesirable publicity in gaining information, advice or help of whatever sort they may be in need. As the services of the Centre are arranged to serve the family in its leisure, use of them incurs no loss of time, nor financial loss concurrent with loss of time.

3. Since the Staff of the Centre come automatically into possession of exact information about each member of the family, the main content of which is necessary for the carrying out of any of the above activities, no time is lost in difficulties of approach, nor in gaining the confidence of families, before appropriate measures can be taken to meet adverse circumstances as they arise. More important still, the Staff often being in a position to know of the needs of the members *before* they themselves become aware of them, can throw a 'spotlight' of knowledge on to the path ahead, while concurrently providing raw material the use of which is likely to further the development of the family; and where health is absent, the Centre may be able to forestall and prevent the occurrence of disease and certainly prevent its chronicity.

4. The very existence of the Centre affords an influence operating continuously to change for its member-families the environment out of which the majority of their disorders have sprung. *Its power to counter disorder is therefore radical.*

5. Where the Centre has been obliged to operate as a corrective agency, its member-families, on being relieved of disorder which tends inevitably to separate or encyst them in society, find themselves already embedded in a cultural milieu into which their re-absorption is easy and natural.

6. The Centre is *statesmanlike* in constitution, for through its operation it enhances the capacity of its member-families to acquire knowledge and experience and hence to take responsible action *for themselves*.

7. The Centre is *basically economic* in operation, for it cultivates human potentiality and forestalls disorder which hinders

expression of this potentiality. It is cultural, not curative; radical, not palliative in method.

8. The Centre, *in the presence of an adequate wage* for the worker, is financially equitable and well-founded. It proffers goods and services elsewhere unobtainable by its members, the consumption of which is highly desirable alike to the members and for the community; and the purchase of which affords an opportunity for the exercise of responsible and progressively discriminative action on the part of the family and the individuals who go to make it up. The Centre being so constituted as ultimately to be sustainable by membership subscriptions,[1] is in strong contrast to the prevailing trend of affairs in which the State or charitable societies themselves assume for the public the responsibility for spending, and for which function these agencies are increasingly driven to set up extensive, dissociated administrations, all too often in an *ad hoc* fashion. By thus shouldering the responsibility of the individual, these provisions are unwittingly depriving the family of one of the primary chances for expression of the health and virility of its members. The Centre—one single organisation—meets in the course of its ordinary routine the needs now partially met by a large and growing number of specialist organisations, many of them overlapping and none of them accurately articulated one with another.

Throughout this narrative, we have seen how the Centre organisation fulfils in abundant measure the family's needs for knowledge and for action in all the phases of its growth: as for example in ante-natal care and infant welfare, in bio-chemical, medical and social inspection of the school child, in the health overhaul and vocational guidance of the adolescent, in education of the parents; and above all in the opportunities it offers to the young couple at the time of courtship and mating. We have also seen the advantageous position in which the biologist is placed for carrying out these duties where periodic health overhaul of the whole family is the basis of his or her approach to every individual; where opportunity to see the individual in action in the freedom of his leisure is added to knowledge gained in the consulting room; and where the instruments for the promotion of health are continuously at the disposal of both the biologist-

[1] See Appendix IX. *Financial and Administrative.*

cultivator and the family. It is incidental, but none the less striking for that, that these conditions, necessary for the cultivation of Health, fulfil abundantly the optimum conditions desired by the therapist—medical, social and psychological alike.

In illustration of this administrative issue let us take as an example Child Welfare. In the Centre we cannot be said to do 'Child Welfare', for as we have described in chapter IX, we approach the child through the family, both mother and father being implicated in all action vis-a-vis the infant.

Now, useful as the Child Welfare Service has proved for the prevention of sickness and mortality, it is at a hopeless disadvantage for the cultivation of health. Not only does it deal with the infant as an isolated unit, but in many instances it is not even in contact with the mother during, nor is charged with the care of, the most crucial phases of differentiation, namely early foetal life and the first few days after birth. Its primary aim—that of the cultivation of health—may thus well be a lost cause before it can begin!

There will of course always be the sick child, for whom clinics will always be a necessity. But the Infant 'Clinic' as it at present exists is an anachronism. It is not a clinic for the sick—for its doctors, and with some justification, are not permitted to treat the infant if sick, but must refer it to a clinician. Nor is it a 'health organisation', for it has no instruments for the cultivation of health, and is content to work under the disability of dealing with the infant as an entity dissociated from the natural mechanism for nurture—the family. How could the horticulturalist cultivate his plant if his procedure were strictly limited to the care of the bud as it opened? All he could do would be to apply insecticides and powders to preserve it from the depredations of its enemies: that is, to cure its infestations. The position of the Infant Welfare worker is but little better.

In all too many cases the Infant Welfare Clinics have no contact at all with the family before the birth of the child, and have to rely on house to house visits of the Health Visitor to ensure the mother's attendance. Where Infant Welfare and Maternity services are carried on as separate organisations with separate staff, these visits are made for the first time *after* birth of the child.[1] The mother comes at her leisure—three to six

[1] The law provides that they shall not be made till the midwife leaves; i.e. after the 14th day!

weeks later—when the first steps in the 'birth weaning' have all too often become but a series of blind stumbles, ending, often before arrival at the Clinic, in abandonment of breast-feeding and in the gathering chronicity of those puerperal ailments of the mother regarded as minor and therefore all too often disregarded.

It must be noted that if at the 'nesting' period the dwelling is to be kept inviolate, which, as we have shown, is instinctive and necessary for function, all educative factors concerning the hygiene of the nest, etc., must be undertaken *before* birth; for, whether well or badly wrought, from the biological point of view the nest can only be made worse by intrusion of that which is foreign and non-specific, when it has become the seat of functioning. So that any form of organised visitation of the family by strangers during these early weeks is to be deplored. A little boy's finger in a bird's nest may lead to the abandonment of six eggs and the making of a new nest and a new effort at egg laying. Too early handling of a rat's young may lead her to eat them. The human mother so interfered with is liable to eat her child—spiritually.

It is well known and universally deplored that the Infant Welfare Organisation has difficulty in retaining its hold on the infant that is not sick; while all deplore the fact that attendances fall away by the end of the first year and the growing child is then lost sight of.[1] The scope of the organisation and its appeal are too limited.

There is one more point: and one no less important. All organisations to maintain their appeal and value to the community must allow of the progressive development of the craft of those employed within them. What of the Infant Welfare practitioner? He or she, trained in Hospital upon the sick child, has in the majority of cases never been in contact with a child *not sick* till taking up Welfare work. Thus, initially without special experience, it is at her job that she must learn all she knows. But there, armed with nothing but a stethescope and a weighing machine, she is cut off from all the scientific instruments

[1] Although the attendances of infants at Welfare Clinics appear high, the figures are in fact misleading, for they are arrived at by dividing the total attendances of all infants by the number of infants attending. Thus where some attend regularly, many may have attended but once during the whole period of infancy.

of her profession and is bereft of a laboratory for routine or research purposes—a primary essential to all scientific practice. She is so overburdened with work that she must attend almost exclusively to the most disordered and has but little time to give to cultivation of the healthy; she has no commission to devise apparatus or instruments proper to the cultivation of health, and in many cases has not even a Nursery at her disposal and under her scientific direction, to which to refer either infants or mothers needing education or scope for development—even as a preventive measure for impending difficulties clearly foreseen. The terms of reference of her post forbid her to treat the mother, bound up though the mother is with the infant, and—unless out of hours—she has no opportunity of knowing the father.

Last of all—specialist though she is by calling in Infant Welfare—she has no place in nor authority over the care of the infant in the Maternity Home, which is under the control of an obstetrician who is not 'a specialist' in infant care. 'Indeed, so poor is the co-ordination of medical work in this field that the Child Welfare Officer does not *as a routine* receive the courtesy of official information as to what has befallen either infant or mother during confinement, indissolubly linked though they are at this period.

Thus bereft of suitable equipment and circumstances, and equally shorn of a proper professional status, how can we expect progress from the doctor in this field? As for research, the possibilities are indeed remote. The organisation—as Child Welfare workers, medical and lay, fully realise—is not comprehensive enough to allow of rational handling of the work it is designed to do—namely the *cultivation of health.* A club for families such as the Centre, on the other hand, fulfils each need of all concerned with the care of the infant.

Let us take one or two other examples showing how the service rendered by some of the special agencies listed above are met by the Centre organisation. The conduct of Child Guidance affords a good example. In the Pioneer Health Centre there automatically accumulate in the dossier of each child :—

(1) Health records of the individual which include :—

(a) ante-natal and post-natal records up to school age ;

(b) periodic health overhaul at not more than six-monthly intervals after he reaches school;

(c) running observations on his physical, mental and social aptitudes as they have developed spontaneously in the freedom of his leisure.

(2) Records of his family (parents and brothers and sisters) from the medical, social and functional aspects.

The basic information necessary is thus already at hand, and *from the previous personal observations* of the person whose duty it is to deal with the child. Still more important perhaps, once the nature of the disorder is diagnosed, there are at hand instruments and social situations, all of which are at the disposal of the biologist for the early rehabilitation of the child; and it must not be forgotten that this includes a situation in which the rest of the family take their part; they are therefore as open to influence as is the child himself. Lastly, instead of having to congregate a difficult child with other difficult children in the process of being cured, in the Centre the child is already in his place among a preponderance of normal children and in a mixed society which includes people of all ages—in itself a health promoting factor, inducive of social 'order'.

The first person to admit the need for (1) intimate contact with the family of the difficult child, and (2) a means of modifying the environmental conditions of that family, is the Child Guidance practitioner. Apart from the Centre these necessary conditions have however proved to be among those most difficult to achieve.

Let us take as another illustration of a provision recognised by the clinician as necessary but hitherto inadequately met: namely the 'follow-up' after all sorts of institutional treatment. On discharge from costly treatment of a highly efficient and elaborate technical order such as surgical operation or the care of fevers, the patient, debilitated from sickness and inaction, is returned home. Illness has probably drained the family purse, anxiety has worn down its resistance. It is into such a home atmosphere that the invalid is returned, cured of his illness, but nevertheless still a victim of the debility resulting from it.

As a member of the Centre he straightway receives, not a medical examination, but a health overhaul. His depleted reserves are attended to, his physiological capacity assessed. Thus armed with knowledge of his condition, he now finds at his disposal the instruments of the Centre and its society to draw upon for his convalescence. In this way the skill that has been so lavishly provided for the care of his illness in Hospital is enabled to bear maximum fruit in a quick recovery of health. Meanwhile the spirits of the patient and of his family are continuously sustained and refreshed, and the retreat from society that comes so commonly with ill health in the family, is avoided.

Many a child returns from the Fever Hospital debilitated, only to crush its nose against the window pane of the front parlour for weeks of ineffectual convalescence. Its parents all too often are in complete ignorance as to what happened to it while in Hospital, its reserves are depleted and its minor ailments go unattended. Many a heart case or healing fracture remains stationary for weeks, lack of knowledge having engendered—in the individual or in his parents—the fear of bringing a once injured organ or limb into active use on return home. How invaluable, for example, it is at such times to have at hand a swimming bath—and under supervision—for the after-care of this type of case.

Preventive measures in general fall no less easily within the Centre organisation. Take for example immunisation against infections. In the early months of the Centre's work, measles broke out in a neighbouring school. A mother who had a child at that school asked for an appointment to see the doctor for a few minutes. She had heard that children could now be protected against measles and wanted to know if her boy could 'be done', and if it would be wise to have her baby done too. Could the Centre do it for her ? The medical staff had been waiting for just such a question. Willingly. An appointment was made and the mother brought her two children. Afterwards she had tea in the Cafeteria and there friends gathered round her to hear what she had been doing upstairs. Could they have their children protected too ? A stream of enquiries reaches the Receptionist. If they have young children at the same school, appointments are made for them ; if not, they are told that since this immunisation is not of long duration, it is inadvisable to have it done unless their children are directly exposed to infection, in which case they should come at once. They understand. Perhaps as they leave the Consulting Room it is time to fetch their babies from the Nursery ; there the gossip spreads ; more mothers hear ;

all begin to want their children protected too if there is any danger. No posters are necessary; no persuasive lectures; only a free seeping of information throughout the society of the Centre and there is stirred a growing sense of the family's responsibility for the nurture and care of its own children. The Centre has aroused an 'appetite' for parental action and has avoided the necessity for persuasion or coercion, the use of which so often engenders reaction and militate against progress.

Immunisation against small pox, diphtheria, whooping cough, scarlet fever, septic infections and many anaphylactic conditions are all carried out in the Centre, the desire for them arising in the same topical way or through knowledge spread by gossip among the members. It should be noted in this connection that where the immunising procedure is a well-established one, the cost of the material is always made known to the family (for the purposes of education) but the cost of the service is regarded as covered by the family subscription; the net cost of the *material* is paid by the family—and ungrudgingly. In the case of vaccination or immunisation against diphtheria, all our members were in a position to choose whether it should be done in the Centre or, without cost to them by the public services. In the case of diphtheria, the total cost of immunisation with subsequent Schick testing was 2s. 6d. per child when shared by several families. The cost of vaccination was 9d. for one infant, 4½d. when shared by two. To reduce this cost to the minimum it is thus in many cases necessary that groups of families should share the material. The finding of friends, or the meeting of mothers not known before, to form a group of a suitable size, is in itself a factor helping to spread both the knowledge of and the desire for immunisation. In the Centre every incident like this affords a possibility of a new social contact, thus tending to social integration, and to social education.

There are among the objects covered by the list given in the early part of this chapter, several of which impinge on the married woman and mother. There is nowhere outside the Centre any cultural organisation—with the exception of the Women's Institutes and Townswomen's Guilds—which is centred about *her*, to encourage and foster her education as a home builder. Still less are there any centred on the *family* as the natural instrument of growth and development. The Centre is the only organisation, so far as we know, which operates exclusively on the family, and which does so acting at all points topically for the education of the family as a whole—father, mother and children simultaneously.

Usually to enjoy facilities such as the Centre offers, the individuals of a family have to scatter in different directions over the town, while almost all are too far away for the mother to use them with any regularity. But where, as in the Centre, so many facilities are centralised, all are brought within reach of the whole family. It is necessary to stress the fact that the Centre affords a local medium of society easily penetrated by the mother, for only in such a way can she day by day draw knowledge and experience to knead into the life of the home as the primitive woman kneaded the dough that nourished her family. Unless such experience and knowledge can reach the mother, the experience of each individual of the family cannot be transmuted into and become an integral and live part of the home.

In a household where there are young children, the rhythm of the mother's day is set unalterably by school attendance with its consequent journeys to and fro, alternating with meal times, preparation of the children for bed, etc. Into this framework she fits her domestic work and within it finds her necessarily scattered leisure. When the father's working hours are such that his return to the evening meal, even at a late hour, is regular, the family have the advantage of some leisure together. But often there are conditions over which the family have no control; for example in many cases the man's work is in shifts, changing every week, so that the varying times alter his meals and sleeping periods and so fragment the working programme and curtail the leisure hours of the whole family. But an hour snatched here and there is available, so that only some nearby provision, that is to say some provision that lies in the path of her daily comings and goings, can meet the housewife's needs. Yet the less time there is, how much the more necessary is some local provision for refreshment and for re-creation of body, mind and spirit. The Centre fulfils this need, however unusual the circumstances of the daily routine of each family.

The effect upon the members of the family of this centralisation of the family life around the mother and her welding or digesting influence upon material so garnered into the home, is obvious to us as observers in the Centre. It is equally obvious that in those relatively few cases where the mother cannot or does not use the Centre herself, the activities of the rest of that family in the building fail to become integrated into a social whole. It follows that nothing but some *local* provision serving for the daily leisure activities of the whole family can produce the integration of modern society, and that it must be centred within the ambit of the mother's excursion—for as we have seen, though it is parenthood which creates, *it is womanhood that must gestate and bring that which is created to viability.*

CHAPTER XVI

A COMMUNITY GROWS

In the Centre we have not been looking at men at work, infants at a clinic, children segregated at a play-centre or at school, adolescents beginning to earn their living or gathered into Boys' or Girls' Clubs—nor indeed at any isolated group or class of individual as commonly envisaged for the purposes of present day administration. We have looked for evidence of function in that long pulsating stream of livingness in which human families fulfil their cycle of development; where husband and wife are seen as one, united in parenthood, and where the child, not regarded merely as an isolated individual, is seen as a new 'limb' or differentiating organ, arising and acting within the unity of the family—concrete and tangible evidence and sensitive indicator of the development and functioning of the whole family organism.

It should not then surprise us that no list of the various separate ways in which the Centre meets the needs of the family will reveal its full significance for that family. It is of course true that the chances for knowledge and for action there made available are not otherwise wholly denied to the man-in-the-street. Any young family who takes the trouble and the time, can seek out a Welfare Centre, a good doctor, a Public Library, a swimming bath, etc., and can thus find for themselves, if they know what they want, a good deal of what is to be found in the Centre.

This phrase—'*if they know what they want*'—gives us a clue as to how it is that what they find in the Centre comes to serve as food for their growth. Knowledge of how to go about things is gained above all from living in an environment in which the example of competent action is all pervasive. The handicap of the scholarship boy is not his poverty—that can be met and provided for; it is the cultural poverty of his home in the biological sense we give to 'home'. A young family living in the social isolation of modern urban society that we have described, may be alive to its situation, but has little ability to formulate its needs or to go out in search of them—as the condition of these families when they join the Centre so clearly demonstrates. Still less can such a family conceive of its future needs, nor plan for them. How can a young couple know that anything even so commonplace as the use of a badminton court or a swimming bath, or membership of a dramatic group, will prove invaluable to its social integration in the near future—in a month's time, or in a year's time—and that either of these things might turn out to be an important factor, for instance, in bringing about the smooth weaning of its first-born, saving them and their future child from the psychological thraldom of 'skirtboundness' with its consequences for them all ?

In health, the future *grows* out of the present, not as a plan, but spontaneously as the family grows in mutual action with an environment that also is alive and growing. In order that the potentialities of a family may develop, it is essential that—like the ovum bathed in rich nutritive fluids already in some measure familiarised for it—it should be in a position to select those things that it needs as it needs them ; not that it should have to go out, and *imagining* its necessities out of void, seek from among wholly unfamiliar material those ingredients that are essential to it ; nor that, remaining in ignorance of the possibilities, it should have to rely either on the patronage of the 'haves' or on the non-specific doles of a bureaucracy however benevolent, to provide its needs *ready-made*. Nor indeed that, once devitalised and its home shrunken, it should have to be spoon-fed with those items in which the therapist—medical, social or psychological—can recognise it to be lacking, and must perforce at this juncture administer *as drugs*.

All the chances offered by the Centre, chiefly in the form of raw material as we have seen, are there to be woven by the family itself into the contextual fabric of its gathering experience. The setting of the Centre's equipment—instruments, knowledge, people—is such that all action arising from its use can readily and naturally form both the stimulus to and the substantial focus of a growing social life for the family. And by 'social' we do not mean that casual use of the adjective to describe the amorphous aggregate which a man joins as spectator at a football match or cabaret, or when visiting a Road House. We do not use the word to denote a man isolated amidst the concourse of his fellows, but rather to describe that unique faculty in man, whereby in mixing with his fellows he not only extends his awareness of the world around him, but, in mutual action and in continuity of friendliness, as *an integrated organ of an organismal society*, fulfils and sustains himself and his own family, while at the same time enriching his environment—in which his fellows share.

This state of relatedness of the family in society—as of the individual in the family—implies some structure for its growth—its own 'zone of mutuality'. The Centre fulfils its purpose because by reason of its nature, and of its relationship to the family, it constitutes what might be likened to just such an 'interfacial membrane' in society; an active potent surface across which material can freely pass for utilisation on both sides. On one side of the membrane is the ethnological unity of the family made up of its individuals, parents and children; and on the other side of the membrane the ethnological unity of the Centre society as a whole, made up of the totality of its family membership.

We see the Centre emerge, then, as a new factor in the environment of each family. The family comes in, joins, and begins to exercise its faculties on what it finds there. So a fragment of the Centre becomes included in the family flow. The home has engulfed, familiarised and digested its morsel. From that moment the Centre is something 'alive' within the body of that family; the family is something alive within the body of the Centre. At this point, for the family that has appropriated it, the Centre disappears as an *institution* and becomes part of its own live and familiar environment—part of its *home*.

Meanwhile, the Centre's own development begins to be guided by that member-family, as in pregnancy we saw the foetus guide the development of its mother. So the two together, member-family and Centre, form a *zone of mutuality*—as it were a social placenta—in the living body of society. A small bit of society itself is now in process of organismal development.

And here perhaps we have an inkling of what that word 'community' which has such charm for us, may imply. Certainly those who first sought to give the word a modern meaning did not imply the commonality of non-specific association for which it is so frequently used at the present time. 'Community' is not formed merely by the aggregation of persons assembled for the convenience of sustaining some ulterior purpose, as in a housing estate connected with a single industry; not by the aggregation of individuals kept in contiguity by the compulsion of necessity, as in 'special areas' wrecked by unemployment; nor held together, as in some Social Settlements, by the doubtful adhesive of persuasion; nor indeed meeting the needs of war time as in 'Communal Feeding', 'Communal Nurseries'. Its characteristic is that it is the result of a *natural functional organisation in society*, which brings its *own* intrinsic impetus to ordered growth and development. In our understanding, 'community' is built up of *homes* linked with *society* through a functional zone of mutuality. As it grows, in mutuality of synthesis *it determines its own* anatomy and physiology, according to biological law. A community is thus a specific 'organ' of the body of Society and is formed of living and growing cells—the homes of which it is composed.

The Peckham Experiment has evolved a *Health* Centre, so far as we are aware the only one in the world. It is not that assemblage of Clinics conveniently congregated for the carrying out of medical desiderata, such as early diagnosis, minor therapy, prevention of disease, etc., which it is becoming fashionable to call a "Health" Centre. It is a locus in society from which the cultivation of the family—living cell or *unit* of society—can proceed, and from which the family sustained in its own growth and development, can spontaneously evolve as part of a larger whole—a live organismal society.

Let us then, with this in mind, as silent observers spend a day in the Centre, and watch the families as each dips for nectar in the flower of his choice.

It is 2 o'clock on a warm day in early summer; the Staff and students have arrived and the Centre has just opened. Three men have taken swimming tickets and two have gone upstairs for a game of billiards. We know them well; two are bus drivers, two night drivers of lorries and one is having his week's holiday. One or two school children with an extra half-holiday are already in the swimming bath and a boy of 15 goes up to have a game of billiards with his father on the men's table.

During the afternoon, up till 4 o'clock, a steady stream of young women with their babies come in, filling the long corridor with prams. They disperse, first to the Nursery, to the Physiological Department or to the Babies' consulting room. By

3.15 the whole building is alive with activity. All the babies have gone to the Nursery ; the women's "Keep Fit" has started ; three mothers are dressmaking and using the sewing machine. In the consulting rooms, the laboratory, the babies' consulting room, there have already been ten or twelve mothers keeping their appointments with the biologist or the bio-chemist, with their twelve or sixteen babies and toddlers. In the theatre where the badminton for beginners is going on, some of the players of varying skill who play on other afternoons in the week, have come to show the new members how to play. Here again there are one or two husbands with their wives ; the men are having their annual holiday, and they mean to have mastered the rudiments of the game before the week is out. Later there is to be tea on the stage—a special party—bringing to the fore a new feature in the form of domestic skill in cooking and preparing tea and doing the honours as hostesses for the learners' club.

A group of mothers with some visitors are sitting in the main hall within view of the gymnasium into which a group of three and four year olds have just come. The student in charge looks up and smiles at them as she and the children set about getting the apparatus ready—putting out the forms and the horse so that the children can climb and slide and balance. A mother passing the gym on the way to the Nursery with her youngest, stops to watch and to discuss with the student how long it will be before Bill can join his sister, and then she takes him to see the even younger children in the learners' pool where they are sitting on the steps of the gradually filling bath, jumping about, splashing. She leaves the child in the Nursery, and then goes up to the Cafeteria kitchen, for it is her turn to help in preparing tea for the Nursery children.

And so it comes to 3.30. The busiest housewife is finished, changed, and has come round to the Centre for an hour or two. The knitters in the Cafeteria put down their work, pick up the bundles of towel and swim suit and join those more energetic members who have done their "Keep Fit" first. Soon there advances along the side of the bath a surprisingly large number of young, middle aged and oldish women. But few of these were swimmers when they joined. The learners sit on the edge at the shallow end ; many, bolder by virtue of only a few months or weeks of having learnt to swim, jump from the side or even dive off the low spring board. The few husbands who are free are there to join in the fun. Busy tuition goes on. By 3.45 the activity that at 3 o'clock was dispersed throughout the building has become concentrated in the gleaming pool and in the toddlers' pool downstairs.

From 4 o'clock onwards the swimmers are coming into the Cafeteria, glowing and invigorated after their effort. There is a brighter eye, a greater sociability with staff and neighbours. A warm response greets the school children that flock in after 4 o'clock, and the quiet buzz of grown-up conversation changes to a higher note as the children find their parents and their excited chatter bursts forth. The mother they had left at 1.30 a busy housewife at the kitchen sink, is now changed and looking very attractive among her group of friends.

The children do not stay long—perhaps only to dump their school bags. We have already described how they quickly disperse throughout the building. So the Cafeteria settles back into its quiet predominantly female talk. A little after 5 o'clock a few more fathers have come in straight from work to have a cup of tea with their wives.

By 6 o'clock the mothers with the babies and younger children have gone. It is getting near bed time, and the husband's meal must be ready when he returns. We feel the pulse of family life regulating the Centre, for, with the departure of the 9-12 year olds around 7 o'clock, the building becomes very quiet and rather empty. The staff use this time for their own supper, and there are usually also in the Cafeteria a few family groups who are having a supper party.

Now into this relative quiet there comes somebody new—somebody who has not been seen before to-day. It is the young adolescent, later from school or work than his younger brother or sister. These older boys and girls come into our field of observation as they ask for their tickets for billiards, dancing or swimming, most of them quite unaware of how revealing their actions are of their sexual and social development. Much of the description of adolescent behaviour in chapter XI is based on the observations made by the staff during this time after the younger children have left. For these boys and girls to be allowed to stay so late is a newly won privilege of "growing up"—a recognition by their parents of the children's budding adolescence. The boys, but especially the girls, seem to enjoy walking round the emptiness of the building at this time, as though, conscious of their immaturity, they were using it as a trial trip --feeling themselves into what will soon be their own adult society.

This is the lull before the crescendo of the day's activity, which starts soon after 7.30 and reaches its apex about 9.30. The young men and women come first; the older man who, having had his tea and washed, is rarely ready till after 8 o'clock, comes later with his wife; a stream pours in, until by 9 o'clock there are between 500 and 1,000 people and the building hums with the activities of men and women of all ages from 14 to 91.[1]

Billiards, table tennis and darts are all in full swing and the groups around talk and wait their turn. The swimming bath is full, all the individual hot baths are in use, men are boxing in the gym. There is a whist drive upstairs, and the "Tuppeny Hop" in the adjoining room. In the theatre there is a dramatic rehearsal on the stage, while the badminton "A" group have a match in the auditorium. There are 30 people in the various sections of the "medical" block, where consultation and examinations are continuous. Small crowds at the big glass windows on the main floor watch the badminton and the boxing. There is a band rehearsal in a small top room. The "wireless boys" oblivious to all who pass through, control the broadcasting for the building. A small discussion group, whose subject is "Conscription" meets in another top room, into which all who go to the Physiological Department can see as they pass. In the crowded Cafeteria the Concert Party are selling their tickets for their show next week, and in the small office off the Cafeteria there is a committee of 7 men and 4 women planning the next Sunday programme—for Sunday is a big day when members are entirely responsible for the control and finance of the Centre's activities. In the Cafeteria members stream past the desk, paying their weekly subscriptions. Some wives at the counter are buying vegetables or milk from the Centre's farm; many are helping themselves to coffee or beer, sandwiches or salads at the self-service counter.

[1] See Appendix X; tables showing times of entry of members into the Centre.

Often large parties of visitors come; foreigners who want to see English urban life, scientists interested in some aspect of the work, social workers keen to 'get things going'. All move through the building as part of the company, arousing no comment, causing no cessation of the activity, finding in the members a readiness to talk, to listen or to discuss, if and when occasion arises. Or, there may be a visit from a party of from 25 to 30 medical students from one of the Teaching Hospitals taking their Hygiene course. A lecture first at 5.30; a visit to the laboratory; a swim; supper in the Cafeteria and a glass of beer; and perhaps a talk with some of the members.

In and about this hub of activity all are free to wander. This unrestricted circulatory movement is part of the essential mechanism whereby the activities of the individuals contribute to the integration of the family and of society. The members, the staff, the accompanied visitor, meet and stop and talk. The material for their conversation is there before their eyes, in the innumerable activities going on all around.

Each of course goes round for his own purpose. The young men and girls are usually looking for somebody and at the same time closely watching what is afoot with their elders. But the older people, particularly the men, will stroll round, and mixed with their personal enjoyment is a dawning appreciation of the significance of what they see and its contribution to family life. How often the older husband and wife will remark, as they sit at a table having a quiet talk—"If only we'd had this when we were starting out, what a difference it would have made".

This appreciation of the significance of the Centre has often been expressed by the older men and women in what they have done. Much of the cooking for parties and all the preparations are made by the wives, and there are remarkable accomplishments by the men, the manufacture of a whole outfit of fair booths for the Coronation celebrations, the construction of the stage lighting outfit—a professional job—the stewardship of the country camp, the conversion of an oasthouse into a hostel equipped for holiday use for families. All this work requiring the skill of the bricklayer, the electrician, the carpenter, the house decorator, the wireless engineer, the blacksmith, has been undertaken with a verve and a generosity of time that is the members' way of expressing their comprehension of the meaning of the Centre and their anxiety to play some part in this society of families. There is a quite unconscious mutuality of effort here, contributing to a full social life.

In families maturing in these circumstances, when the time comes for the adolescent to go further afield, to leave the nest, the parents are not just discarded, left behind, but continue with their own established place in the community. The family may be reduced in size by the young people's departure to form their own new families, but its excursive possibilities do not wholly disappear nor does the 'home' built by parents and children together necessarily contract and shrivel. Contrast this picture with that of the lonely old couple who, left behind in their foreshortened growth, still cling to their children and remain utterly dependant upon them for any social excursion; or who wait at home for that visitation of charity—a friend who calls out of kindness.

As is to be expected in a building reflecting so sensitively the life of its member-families, as well as the ebb and flow of daily use there gradually emerges the weekly rhythm of family life with its less frequent festive and serious occasions. Each day of the week has its own characteristic savour—and one gets to know who one can expect to meet. Then there is the festive party at Christmas, the Sunday morning swim and the serious Sunday evening lecture, the special character of a Bank Holiday —one of the rare occasions when whole families can spend their day together in the Centre and when those who have gone further afield call in on their return in the evening for supper, or a drink and a chat.

Here then is a chapter in the biography of the *FAMILY* functioning through a live and growing *HOME*. From the nucleus of parenthood a family has grown; its 'limbs' having gone far and wide into their diverse environment, engendering a functional organisation of infinite delicacy and sensibility, intricate and far-reaching in its contacts with its neighbours. Family by family, each thus inhabiting its protean home, makes its way among its neighbours, encountering all, acknowledging many, welcoming where it wishes—until its environment itself becomes an organised community of families. As each family grows, so its society grows with it.

Such is no mere association of groups of people each with an objective. We give a description of mutual subjective synthesis of family and environment alike, giving life and form to Society. Just as our own body is made up of cells, so this community is made up of homes, whose 'interfacial surfaces' are absorbing material and experience that is in circulation throughout the whole social body, that body being modified the while by the synthesis of each and all of its component homes.

So in the midst of social disintegration here there is beginning to appear a nucleus of Society the structure of which is neither 'planned' nor 're-constructed' but *living*; that is to say growing, developing, differentiating, as the result of the mutual synthesis of its component cells—its *homes*.

And so we return to our starting point, Function. Like the physicist with his unit of construction—the atom—or the physiologist with his unit—the cell—we, beginning with the 'family' as our unit, arrive at the delineation of a field through which is expressed the specific functional action-pattern of that unit. The field of function, the 'home' is the biologist's functional 'cell'. It is from home to home, cell to cell, through the function of parenthood, that biological Energy is transmitted. The individual is but an evolutional dead-end—a cul-de-sac of biological Energy seeking canalisation.

It is on the 'cellular' concept of Schwann that exact knowledge of the physiology and pathology of the human body has been based. Perhaps this 'cellular' concept which we have taken as our premise—of the family-functioning-through-its home—will prove to be the basis for the growth of a *Science* of the *Living* structure of Society.

THE NATIONAL HEALTH TRUST.

If while reading this book you have had a window in your mind opened—learnt something of value for your own work—recall that you are indebted to all those, rich and poor, who have given and are giving their money and time to make this experiment possible at all.

You can discharge your debt and support this research by joining the National Health Trust.

The object of the Trust is to unite and strengthen the position of those who believe that, important though it is to treat and to prevent disease, it is even more important to CULTIVATE HEALTH.

Membership of the Trust will keep you in touch with the progress and development of the work in this new field.

MEMBERSHIP.

I desire to become a MEMBER of the National Health Trust and I will subscribe 10s., £1, 30s., £2, £3, £4, £5, etc., per annum.

Signature.................................. Mr., Mrs., Miss

Address..

....

Bankers Order :

To the..Bank

....

PAY to the NATIONAL HEALTH TRUST through the MIDLAND BANK, LTD., BLACKFRIARS, LONDON, E.C., the sum of

£ : : on this date........................, 19 ,

and on January 1st of each subsequent year, until further notice, and debit my account therewith.

Signature ....

Address ....

DONATION.

I desire to give as a donation to the NATIONAL TRUST for the Promotion and Study of HEALTH the sum of

£ : :

Signature ....

Address..

....

[*Alternative*]

A DEED OF COVENANT.

If you sign a Deed of Covenant to pay a subscription for seven years this enables the Trust to obtain a refund of Income Tax year by year, substantially increasing the effective amount. Thus a subscription of £5 0s. 0d. per annum with Income Tax at 9s. in the £ produces a total of £9 1s. 10d. for the Trust.

If you are willing to increase the effective amount of your subscription in this way, please write to :—

The Secretary,
The National Health Trust,
8F, Hyde Park Mansions,
London, N.W.1,

who will send the necessary form for signature.

APPENDICES

		Page
I	Notes on the Building	301
II	Services and Amenities	303
III	Nature of Employment of Members	307
IV	Specimen of Laboratory Records	311
V	Cost of Health Overhaul	313
VI	Prevalence of— (i) Iron Deficiency	315
	(ii) Worms	315
VII	Plans for an Educational Experiment	317
VIII	A Child's Activities	318
IX	Financial and Administrative	320
X	Use of the Centre—(ii) weekly ; (i) daily	323

SKETCH PLANS

The general construction of the building is that of flat slabs on cruciform columns allowing of maximum flexibility of planning: no immovable partitions.

Whole building planned on a grid 18ft. square, varied at either wing to give a 24ft. span for gymnasium and theatre.

All single lines on the plan represent easily removable brise block partitions.

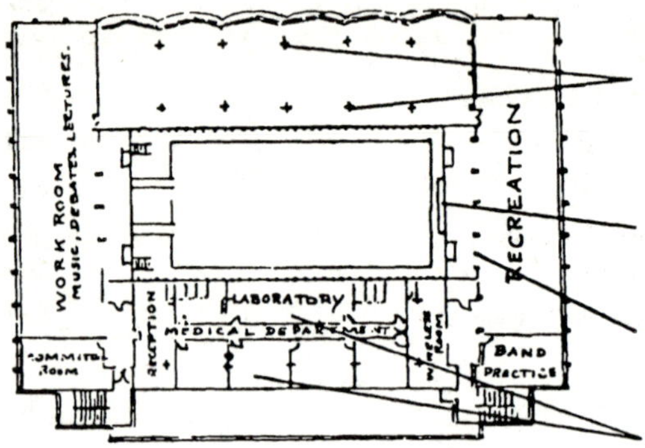

SECOND FLOOR PLAN

Cruciform pillars carrying the concrete floor spaces, and affording conduits for power, water, etc., allow flexibility of internal planning.

Window giving view of swimming bath.

Billiards, table tennis, darts, etc.

There is privacy in the Physiological Department.

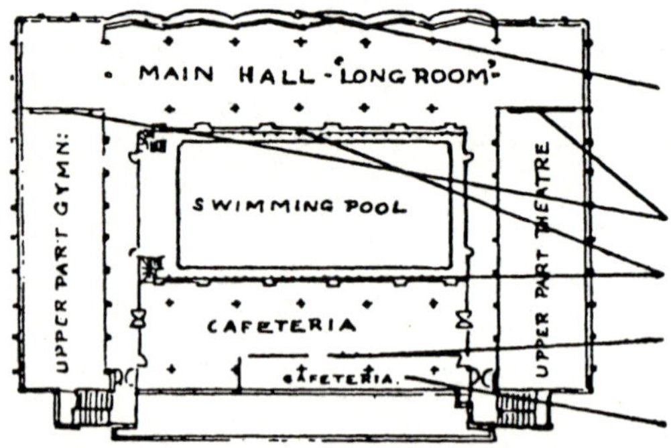

FIRST FLOOR PLAN.

Sliding windows the height of the room can be thrown back in Summer. Bays between the pillars forming natural points of congregation for groups of people.

Windows giving on to gym and theatre.

Windows surrounding bath.

Cafeteria counter and window into kitchen.

Kitchen for preparation of cafeteria meals, including **Nursery teas by mothers.**

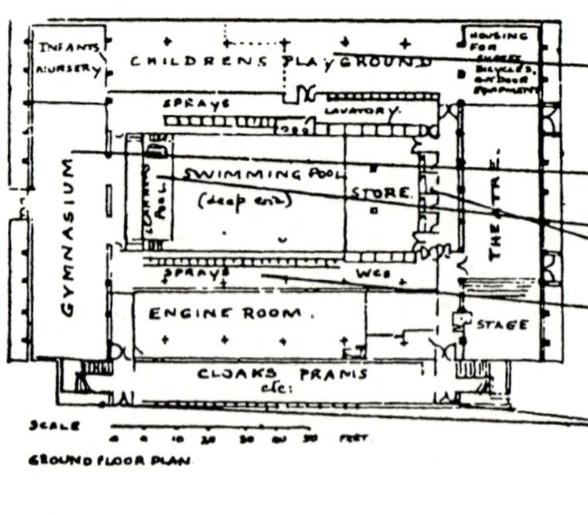

GROUND FLOOR PLAN

Covered playground for children of all ages, with Infants' Nursery located to catch the full afternoon sun.

Gymnasium with easy access to the Nursery, playground, and to the garden.

Infants' and learners' pool.

Slipper baths

Lavatories and spray-chambers for men and women on either side of the bath chamber.

The main entrance is at the back, leaving the front unobstructed for maximum utilisation of sunshine and open aspect.

APPENDIX I

Notes on the Building

The design of the Pioneer Health Centre was brilliantly executed by Sir E. Owen Williams, to fulfil the requirements of the experiment laid down by Dr. Scott Williamson. Built in 1934-5, its total cost, with basic equipment, was £38,000. A description of the building, from the architectural point of view, can be found in the *Architectural Review*, May, 1935.

FLEXIBILITY OF THE BUILDING: We have described in the course of the text that for the Health Centre to fulfil its purpose, the activities of its members cannot be pre-designed but must *grow* spontaneously. Hence the need that the construction of the building should be such as to give opportunity for maximum fluidity and change in its continuously unfolding 'organisation'.

For example, many occasions may arise when it may be desirable that refreshments, as well as being provided in the cafeteria, should be available elsewhere in the building. Apart from a linked train of trolleys which would make this possible, it is desirable that there should be available throughout the building sources of power and of hot and cold water. This necessity was met in the design of the building by conduits running up through the three floors in the interior supporting pillars, making it easy to tap these sources of supply at will.

HEATING: The heating of the building was carried out by a hot water system fed from storage tanks heated by electricity derived from the main at off-peak load. Experience has shown us however that since not all the building is in use at every moment of the day, it is constantly necessary to raise the temperature of one or other section for short periods of approximately one hour. For this reason a system based on a series of focal sources of heat, quickly raised, would have been better suited to this type of organisation. Since heating and lighting is the major maintenance cost of a building of this nature, this point is of very great importance.

BATH CHAMBER: In a Health Centre, where the use of the bath is for the development of health as well as for pleasure, the users should not take exercise in a vapour-laden atmosphere. For this purpose the bath chamber was equipped with an extraction apparatus for removal of water vapour from the surface of the water. This had the added advantage of preventing condensation occurring on the glass surround, thereby obscuring vision.

NOISE: In a building calculated to accommodate 1,000 or more individuals at one time, the elimination of noise is a matter of great importance. Cork was used as a medium to overcome this difficulty; the ceilings being faced with $\frac{3}{8}$ inch spongy cork coated with a colour wash, any concrete wall surfaces—of which there are few—with $\frac{1}{8}$ inch unpolished cork, and all concrete floors with foot square tongued and grooved cork tiles, dressed to give either a dance or a non-skid surface.

For those interested in buildings of this type for the future, four years' experience has shown the completely satisfactory nature of ceiling and floor surfaces. Cork as a wall covering reduces the noise factor, but while admirably suited to the examination department, has proved unsatisfactory in other parts of the building owing to the attractive surface it offers to the penknives of the idle newcomer.

THE GYMNASIUM AND THEATRE FLOORS: Particular mention should be made of the construction of the special floors designed by Dr. Scott Williamson for the gymnasium and the theatre (used also for badminton and for dancing). Across an unscreeded concrete base, $\frac{1}{2}$ inch rubber $\frac{1}{8}$ bore pressure tubing was stretched in parallel lines. The space intervening between the lines of tubing (9—12 inches according to the degree of spring required) was then filled in with cork chips till level with the summit of the tubing. The level surface thus formed was covered with the same tongued and grooved cork squares as the rest of the building.

The effect of this construction was to give a uniform resilience, with a softness and gentle spring not unlike turf, and without dead areas due to joists. This floor surface proved over $4\frac{1}{2}$ years' continuous use to have an excellent wearing capacity.

Services and Amenities Available by the Summer of 1939

Club open :

Mondays to Fridays	2 *p.m. to* 10.30 *p.m.*	For all activities. (Health overhaul Tuesdays to Saturdays).
Saturdays	2 *p.m. to* 11 *p.m.*	
Mondays to Saturdays	7 *a.m. to* 8 *a.m.*	To members of early morning Swimming Club.
Sundays	7 *a.m. to* 9 *a.m.*	For early morning swimming.
Sundays	6 *p.m. to* 10 *p.m.*	October to March inclusive, when there was a special programme arranged by the members who were entirely responsible for running the Centre on Sundays.

SUBSCRIPTION :

1s. a week per family, entitling all children under 16, or still at school, to free use of all equipment, and adults to use of each facility at a small charge. All children of member-families, over 16 and not still at school, (as also all Temporary Members, see p. 230) pay an individual subscription of 6d. a week, and are also entitled to use all equipment for a small charge.

Services rendered, and activities available in return for weekly membership subscription without further payment :

I CONSULTATIVE SERVICES ; by appointment, between 2 and 10 p.m., Tuesdays to Saturdays inclusive :

Periodic overhaul, comprising :
- Laboratory examination,
- Personal examination,
- Family Consultation,

Any special appointment at request of staff or member.

Re-overhaul on discharge from medical care after sickness.

Advice on contraception.

Ante-Natal and Post-Natal care.

Infant care.

Immunisation against infectious fevers, allergy, etc. (service free ; material at cost price).

Parental Consultations.
- On announcement of conception.
- At birth.
- At each successive weaning period.
- At any other time indicated by circumstances.

Vocational Guidance.

Sex instruction to adolescents (private appointment with doctor).

Health overhaul of fiance(e)s of members, whether themselves members or not.

Pre-marital Consultations.

Advisory service, legal, and other.

II GENERAL USE OF THE BUILDING, including the cafeteria and the main social hall with dance floor, at all times when the Centre was open.
Infants' Afternoon Nursery: 2—6 p.m. daily (Sundays excepted), with use of gymnasium and infants' swimming pool by Nursery children. Preparation by Mothers of Nursery teas, on a rota system (educational); charge for teas to cover cost of material only.
Night Nursery: 8—10 p.m., for children under 2.

Activities available to adults at a small charge and free to children, at all times when the Centre is open:

Swimming bath (½ hour, 3d.) Swimming Club, 6d. weekly, 3 swims.
Hot baths (3d.)
Billiard tables (6d.) (no children on adult tables).
Table tennis (1d.)
Darts (1d.)
Cards, chess, draughts, etc. (1d.)
Cricket practice nets, and equipment.
Equipment for boxing, fencing, etc.
Work room with sewing machine, cutting-out table, scissors, iron, fitting mirrors, etc.

Special equipment provided for children, and available for their use without extra charge to the family (see chapter X):

Roller skates
Shinty sticks (for hockey on skates)
Fairy cycles
Bicycles
Trampolin
Badminton rackets
Small billiard table
Table tennis
Cricket equipment for use at practise nets

Balls
Parlour games: chess, draughts, ludo, etc.
Puzzles
Books
Drawing materials
Sewing materials
Typewriters
Tap dancing shoes for learners
Instruments for percussion band

Activities available on weekly basis at regular times, once, twice or three times weekly, afternoons and/or evenings; payments in most cases collected by secretaries of intra-mural clubs:

Badminton, several groups of varying skill (3d.)
Boxing (3d.)
Keep Fit (3d.)
Women's League of Health and Beauty (3d.)
Swimming instruction, Diving instruction, Water polo } 3d. for use of bath
Fencing (3d.)

Dance Club (2d.)
Dancing instruction (2d.)
Tap dancing class (2d.)
Roller skating (3d.)
The Stage, for Dramatic or Concert Party rehearsals (2d. a week to each member of the group. Also bookable by groups of children free)
Discussion Circle, with or without visiting speakers (2d.)

Whist drives	First-Aid classes
Orchestral practice	Woodwork shop
Dance band practice	Various demonstration courses e.g. on cookery, education of the young child, etc.
Wireless Room	
Gramophone and records	

The charge in each case was designed to represent a rental covering the overheads of the part of the building used, as well as the upkeep of the particular equipment and any special costs incurred.

Special Occasions :

Christmas or New Year's Eve Party (as many as 400 people to five course meal followed by dancing and cabaret).

Children's Party at Christmas ; organised by members, who were responsible for all preparations, including refreshments, as well as for running the party. As many as 600 children.

Birthday Party ; running buffet prepared and served by members, cabaret and dancing. As many as 800 people.

Parties organised by various intramural Clubs ; e.g. the Billiards Club, the Darts Club (50—200 individuals).

Performances by Dramatic and Concert Party groups (making a charge for entrance and filling the Theatre for three or more successive performances). Audience 150—200.

Matches with visiting teams from other Clubs.

Outings and expeditions of various sorts.

Family Parties and celebrations, for which a room can be booked.

Occasional entertainments by visiting players, dancers, musicians.

Associated Activities :

THE CENTRE "HOME FARM" at Oakley House, Bromley Common, Kent (7 miles from the Centre), for production of T.T. and attested milk, of fresh vegetables and fruit for sale to members, expectant mothers and young children having priority of purchase.

Planned for the education of the young family in the principles of food and nutrition. See chapters viii, ix.

The Home Farm provides also a playing field available to groups of members for cricket and football.

The Farm is an integral part of the work of the Centre, but up to now has been financed from a separate source.

GREAT SWIFTS HOLIDAY CAMP, Sissinghurst, Kent. Shortly after the Centre opened, through the generosity of the late Col. Victor Cazalet, we were fortunate in being offered the use of a large acreage as a country camp for members of the Centre. This land, which, as well as pasture, included a large wood surrounding a lake and an empty oast house, served as a base for camping activities. The oast house was repaired, redecorated and fitted up gradually by the Centre members who made up week-end working parties for the purpose. Running water was laid on, and gas in cylinders installed for cooking. The camp was used by member-families for their summer holidays, and also for week-ends by some families with their own or borrowed cars, and by the adolescents and younger married people who reach it by bicycle or bus. Sleeping accommodation, either in the Oast House, in the Centre's tents, or on camping sites for the members' own tents, was bookable in advance in the Centre. It was decided by the members themselves that the charge per night made for use of the camp should be a *family* one—irrespective of the size of the family—like the membership subscription of the Centre itself.

The camp has proved a quite invaluable asset in welding links in the social life of the membership.

Nature of Employment of Adult Members of the Centre

This list, drawn from 500 families taken at random, has been compiled to show the great variety of occupations followed. From it may be gained an idea of the range of the social and financial competance of the membership.

Individuals starred are employers of labour owning their own businesses.

Note. The list has been drawn up from the members' own descriptions of their occupations at the time they joined the Centre. While attempting to assess the nature of their work accurately, we hold no responsibility for any deviation from the accepted trade designations.

	male	female
Engineering Industry		
97 *males*, 3 *females*		
Engineer	9	
Engineer's Apprentice	3	
Engineer's Draughtsman	2	
Inspector Mechanic	1	
Foreman Mechanic	2	
Pattern Maker	1	
Plattern Hand	1	
Core Maker	1	
Instrument Maker	6	1
Tool Maker	2	
Engineer's Mechanic	25	1
Ordnance Mechanic	1	
Moulder	1	
Metal Worker	7	1
Metal Polisher	1	
Rivetter	1	
Sound Engineer	1	
Electrical Engineer	2	
Electrical Foreman	2	
Electrician	16	
Cable Mechanic	4	
Armature Winder	1	
Engineer Storekeeper	6	
*Electrical Contractor	1	
Gas Industry		
22 *males*		
Foreman Mechanic	1	
Fitter	11	
Meter Mechanic	1	
Stove Maker	1	
Stoker Trimmer	2	
Clerk	1	
Collector	5	
Building Industry		
57 *males*		
Builder's Foreman	2	
Bricklayer	4	
Tiler	1	
Plasterer	3	
Carpenter	7	
Wood Worker	2	
Wood Machinist	2	
Shop Fitter	2	
Cabinet Maker	1	
French Polisher	1	
Foreman Decorator	1	
Painter and Decorator	7	
Coach Painter	1	
Cellulose Sprayer	1	
Sign Writer	1	
Plumber	5	
Yard Foreman	1	
Scaffolder	2	
Foreman Stockkeeper	1	
Stockkeeper	3	
Quantity Clerk	1	
Builder's Clerk	1	
*Builder and Contractor	3	

	male	female
*Builder's Merchant	2	
*Decorating Contractor	1	
*Sawyer	1	
Printing Industry		
37 *males*, 5 *females*		
Newspress :		
Compositor	5	
Lithographer	1	
Processor and Router	2	
Photo Lithographer	1	
Press Printers	10	
Printer's Assistant	4	
Stereotyper	1	
Machine Minder	2	
Clerks		4
Book Trade :		
Reader	2	
Machine Ruler	1	
Cutter	5	
Warehouseman	2	
Book Folder		1
Book Binder	1	
Transport		
89 *males*		
Railway Workers :		
Station Master	1	
Inspector	1	
Engine Driver	1	
Engine Fireman	1	
Guard	1	
Signalman	1	
Foreman Labourer	1	
Plate Layer	1	
Labourer	1	
Station Clerk	1	
Railway Clerk	4	
Retired Railway Employees	5	
Bus Service :		
Traffic Controller	1	
Driver	12	
Conductor	8	
Traffic Clerk	1	
Cleaners	4	
Canteen Supervisor	1	
Canteen Worker	1	

	male	female
Tramways :		
Driver	3	
Conductor	1	
Retired	2	
Trade Van and Lorry Drivers	20	
Taxi Driver	3	
Chauffeur	2	
*Haulage Contractor	1	
*Garage Proprietor	1	
Mercantile Marine :		
Ship's Officer	1	
Ship's Engineer	1	
Sailors	7	
Clothing Industry		
8 *males*, 26 *females*		
Cap Maker		1
Corsetiere		1
Dressmaker		8
Machinist		9
Mantle Maker		2
Milliner		1
Tailor	4	3
Tie Maker		1
*Boot Manufacturer	2	
*Hosiery Manufacturer	1	
*Shirt Maker	1	
Fancy Goods		
2 *males*, 7 *females*		
Box Cover Maker		1
Brush Case Liner		2
Fancy Leather Worker	1	1
Lampshade Maker		2
Paper Flower Maker		1
Toymaker	1	
Laundry Workers		
1 *male*, 8 *females*		
Manageress		1
Shirt Folder		1
Workers	1	6
Entertainment and Sport		
5 *males*, 4 *females*		
Variety Artist		1
Musician	1	1
Cinema Operator	1	

	male	female
Cinema Attendant		1
Theatrical Properties	1	1
Speedway Rider	1	
*Turf Accountant	1	

Various Crafts

29 *males*, 10 *females*

	male	female
Blacksmith	1	
Bottler	1	
Cooper	1	
Fur Dyer	1	
Gardener	1	
Glassblower	1	
*Glazier	1	
Gold and Silver Refiner	1	
*Hairdresser	2	
Hairdresser	5	4
Laboratory Assistant	1	2
Leatherworker	1	2
Leather Sorter	1	
Mill Overseer	1	
Photographer Printer		1
Tobacco Stripper		1
Timber Measurer	1	
Thermometer Maker	1	
Racquet Stringer	1	
Undertaker	1	
Wireworker	1	
Lighterman	1	
Lockman	1	
Stevedore	3	

Catering and Confectionery Trades

6 *males*, 10 *females*

	male	female
Cake Maker		1
Canteen Steward		1
Canteen Workers		2
Chef	1	
Cook		1
Sugar Boiler	1	
Waiter	1	
Waitresses		4
*Caterer	1	1
*Public House Keeper	2	

Food Trade

18 *males*, 1 *female*

	male	female
Baker	2	
Butcher (Wholesale & Retail)	10	
Crumpet Maker		1
Milkman	1	
*Baker	1	
*Butcher	1	
*Fruiterer (Wholesale)	1	
*Produce Broker	1	
*Vegetable Salesman	1	

Commercial

63 *males*, 24 *females*

	male	female
Advertiser	1	
Artist, Commercial	1	
Commercial Traveller	13	
Commissionaire, Bank	1	
Insurance Assessor	1	
Insurance Agent	5	1
Shop Assistants	26	20
Packer	2	3
Warehouseman	6	
*Newsagent	1	
*Shopkeeper, various	6	

Secretarial and Clerical

63 *males*, 59 *females*

	male	female
Secretary	1	3
Managing Clerk	2	
Law Clerk	1	
Audit Clerk	2	
Auctioneer's Clerk	1	
Time Keeper	1	
Cashier	1	7
Bookkeeper		4
Shorthand Typist		8
Typist		15
Comptometer Operator		1
Insurance Clerk	1	
Dispatch Clerk	1	
Export Clerk	2	
Switchboard Operator		2
Clerks, unclassified	50	19

Professional

6 *males*, 2 *females*

	male	female
Chartered Accountant	1	
Doctor	2	
Minister of Religion	2	
Solicitor	1	
Nurse		2

	male	female
Miscellaneous		
7 males, 2 females		
Artist . .	1	1
Journalist . .	1	
Librarian . .	1	1
Newspaper Representative	1	
Private Detective .	1	
P.T. Instructor .	2	
Services		
43 males, 4 females		
Navy . .	1	
Army . .	2	
Police :		
Sergeant . .	2	
Constable . .	8	
Retired . .	5	
Fire Brigade . .	1	
Post Office :		
Sorter . .	4	
Sorter (retired) .	1	
Postman . .	8	
Post Office Worker .	2	1
Telephone Inspector		1
Civil Service :		
First Division .	1	
Second Division .	6	
Clerk . .		1
Miscellaneous .	1	1
Messenger . .	1	

	male	female
Maintenance		
9 males, 24 females		
Caretaker . .		1
Charwomen & Office Cleaners		17
Cloakroom Attendant	2	
Club Steward .	1	
Domestic Servant .		3
Handyman . .	2	
Housekeeper . .		3
Lift Driver . .	1	
Valet . .	1	
*Window Cleaner .	2	
Casual Workers		
26 males, 12 females		
Labourer .	24	
Factory Hand .		12
Van Boy .	2	
Unemployed .	1	
Old Age Pensioner .	1	
Gainful employment in the house :		
6 females		
Dressmaker . .		2
*Dressmaker . .		1
Miscellaneous .		3
Housewives .		432
Unmarried girls not gainfully employed		2
Student (over 18)	1	

Financial stability of member-families

In *Biologists in Search of Material*, a classification was made indicating the security of occupation of the chief wage-earner of the family. The 500 families the occupations of whose members has been given above, have been analysed on the same basis with the following results :

In *permanent* employment (municipal and civil servants of the lower grades, busmen, tradesmen and employers of labour) . .	204
In *regular* employment, which may be intermittent but also involves overtime pay (e.g. the craftsman attached to a large firm)	242
In *casual* employment	17
Widows and small pensioners	37

It will be seen from these figures that the material under survey in the Peckham Experiment differs in an outstanding manner from that of most social surveys in that it does not represent any socially submerged or financially impoverished populace.

Specimen Laboratory Record chosen at random giving results of first and second Health Overhauls in man of 24

This record shows intermediate check up on the results of liver-iron administration following first overhaul.

Name: Mr. G. W. Date of birth: 30/3/1911. No.: A.0405.

	1st overhaul				*2nd overhaul*
Date	5/7/1935	16/7/35	11/9/35	24/9/35	18/7/36
Vision :— L	S1 6/6				
With glasses R	6/9				
Ishihara test ..					normal
Blood :					
Hb.	85%, 13.6	79%, 12.6	90%, 14.4	100%, 16.0	100%, 16.0
Reds	3,900000	3,100000	5,000000	5,000000	5,000000
Index	1.07	1.27	0.90	1.00	1.00
A.D.	7.60				7.32
Whites	6,000				7,000
Polys.	62%				67%
S. Lymphs	30%				25%
L. Monos	5%				4%
Eosin	3%				3%
Basoph	0				1%
Myelo	0				0
Sugar (80—140) ..	120				114
Urea (22—45) ..	30				32
Uric ac. (1.03—3.5)		1.0			
Calcium (6.0—6.3)			6.5		
Chloride (500—550)				500	
Kahn	(Where indicated)				
Van der Bergh ..					
Coagulation Rate (1)					+1
Bleeding time ..					
Urine :					
Albumin	nil				nil
Sugar	nil				nil
Acetone	nil				nil
Indican	nil				nil
Urobilin	nil				nil
Blood	nil				nil
Urea	1.5%				1.500%
Uric ac.	(in certain individuals for research)				0.050%
Purins					0.055%
Urea : uric ac. ..					30 : 1
Purins : uric ac ..					11 : 10
Microscopical ..	nil				nil
Ph.					6.8
Height	5′ 6″				5′ 6″
Weight					8-12-0
Spirometer	225				265
Temperature ..					98·4

PLESCH TONOSCILLOGRAMME RECORD of arterial tone and pressure of same man on 1st and 2nd overhauls. This recording is taken as part of the routine examination: (1) on arrival in the laboratory.
(2) after the exertion of a spirometer recording.

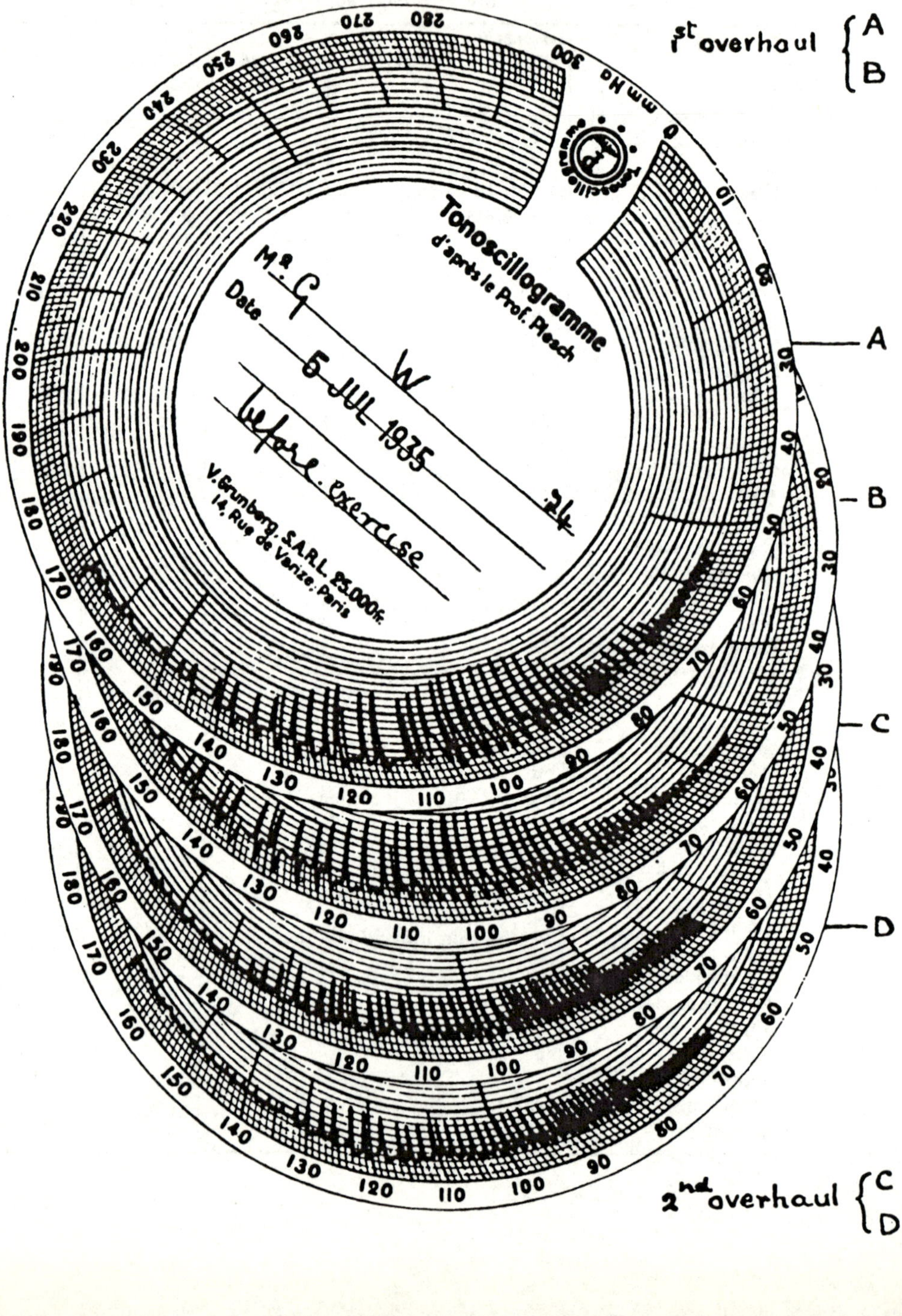

APPENDIX V

The Cost of the Periodic Health Overhaul

The figures for the cost of family overhaul followed by family consultation as conducted in the Pioneer Health Centre were derived from an analysis of expenditure made in 1938. This analysis was made with a view to finding out what, for families, is the essential cost of periodic examination organised on modern methods. From this estimate, therefore, all expenditure on social activities and on research has been excluded. The overhead charges of the Centre have been apportioned according to the space occupied for the examinations.

The membership at the time the analysis was made was 650 families, the staff and set-up for overhaul at that time being adequate for 1,000 families.

The figures derived from this analysis are to be found set out on the following page, the first column representing an approximation of the actual costs, the remaining four columns showing estimated costs with increasing membership.

It must be noted that the data for these calculations is derived from actual expenditure in the Centre where, owing to the research nature of the work, no member of the staff was paid a salary of more than £500. This sum we recognise to be wholly inadequate for any established service. For this reason the costs are set under two separate headings: one indicating the development period with low salaries, the second allowing for salaries at recognised rates (e.g. senior staff raised to not more than £1,000 a year). It will be seen that at present rates an estimated cost of 1/1d. per family per week is reached with a membership of 1,000 families. With a membership of 2,000 families (the number for which the Centre was planned), the estimated cost is again so far reduced that, after making allowance for salaries at an increased rate as well as for additional staff, it remains in the region of 1/2d. per family per week.

In this estimate it was assumed that the overhaul was an annual one, whereas in fact intervals of eighteen months would seem adequate. An allowance of 25% of new members in each year has been made to replace those leaving, and an enrolment talk for these families allowed for. An additional 10% of personal examinations has been allowed for to cover overlapping. All appointments are of 20 minutes duration, with an allowance of 10% for wastage.

Thus, with some considerable experience behind us, we are in a position to state that for 1/2d. per week per family a very exhaustive scientific overhaul including laboratory and consulting room examinations of each individual of the family and a final consultation with the assembled family can be done.

For these figures derived from careful investigation made into the actual costs of the Centre, we are indebted to Mr. Walter Laffan.

Estimated Costs of Periodic Health Overhaul

	Development Period			*Permanent Basis*	
	650 families	1000 families	2000 families	1000 families	2000 families
SALARIES :					
Doctor	£500	£500	£500	£1,000	£1,000
Doctor	500	500	500	1,000	1,000
Doctor	350	350	350	750	750
Doctor	—	—	350	—	350
Doctor	—	—	350	—	350
Biochemist	250	250	250	500	500
Lab. Asst.	—	—	150	—	150
Lab. Cleaner	75	75	75	75	75
Receptionist	—	175	175	200	200
Nurse	200	200	200	225	225
Records	150	150	150	150	150
Records	—	—	150	—	150
Rent (inclusive)	500	500	650	650	750
Materials	100	150	200	200	250
Postages, Telephone and Petty Cash	40	50	90	80	120
TOTAL EXPENDITURE	£2,665	£2,900	£4,140	£4,830	£6,020
Persons overhauled and taking part in family consultations	2,860	4,400	8,800	4,400	8,800
Yearly cost per person	18s. 8d.	13s. 2d.	9s. 5d.	21s. 10d.	13s. 8d.
COST PER FAMILY PER WEEK	1s. 7d.	1s. 1½d.	9½d.	1s. 10d.	1s. 2d.
DOCTORS' HOURS available per year (40 hours per week, 45 weeks per year)	5,400	5,400	9,000	5,400	9,000
Hours taken in examinations	1,500	2,300	4,600	2,300	4,600
% of time taken up	28%	43%	51%	43%	51%
% of time free for research, observation, additional consultations, etc.	72%	57%	49%	57%	49%

(i) The prevalence of iron deficiency

Our point of view and criterion with regard to iron deficiency has already been discussed in *Biologists in Search of Material* (p. 71 et seq.). Using the same criteria the figures below represent the age analysis of iron deficiency (i.e. any fall below 95% of haemoglobin, using the Sahli method and taking 16.8 mg. of iron per 100 cc of blood as standard) found in a total of 4,002 individuals examined.

Age in years	*Males*		Percentage		*Females*	
	No. iron deficient	Total individuals			Total individuals	No. iron deficient
0—5	133	248	53.6	43.5	248	108
6—10	142	200	71.0	60.9	197	120
11—15	114	212	53.8	65.3	199	130
16—20	35	141	24.8	70.2	111	78
21—25	41	135	30.3	63.5	222	141
26—30	44	243	18.1	65.5	226	148
31—35	44	211	20.8	57.9	245	142
36—40	37	188	19.6	57.1	170	97
41—45	29	128	22.7	58.7	126	74
46—50	22	118	18.6	54.9	111	61
51—55	24	86	29.0	41.1	73	30
56—60	24	37	64.8	35.2	54	19
61 and over	17	36	46.6	27.0	37	10
Total	706	1,983	35.6	57.4	2,019	1,158

Included within these figures are 11 cases—2 in men and 8 in women—of clinical anaemia of a severity constituting a major malady.

(ii) Worm infestation

Owing to the laborious nature, both domestic and technical, of examination of the stools for the presence of ova of helminths, and owing to the fact that when the Centre was first opened we did not anticipate the high proportion of infestation later disclosed, only those individuals and families were examined in whom experience came to indicate to us the likelihood of the presence of worms.

The table below represents the age analysis of those individuals suspected of worm infestation, in whom the presence of ova was confirmed by laboratory examination. The figures therefore cannot be taken to represent the total number of worm-infested individuals in our population.

Age in years	*Males* No. with worms	Total individuals	Percentage		*Females* Total individuals	No. with worms
0—5	45	248	18.1	15.8	248	39
6—10	83	200	41.5	23.4	197	46
11—15	55	212	25.9	14.6	199	29
16—20	19	141	13.5	7.2	111	8
21—25	9	135	6.7	5.9	222	13
26—30	7	243	2.8	6.6	226	15
31—35	7	211	3.3	4.9	245	12
36—40	4	188	2.1	4.7	170	8
41—45	3	128	2.3	7.1	126	9
46—50	2	118	1.7	1.8	111	2
51—55	1	86	1.2	2.7	73	2
56—60	—	37	—	—	54	—
61 and over	—	36	—	—	37	—
Total	235	1.983	11.9	9.1	2,019	183

Plans for the First Steps in an Educational Experiment

In the summer of 1939, in conjunction with Dr. Montessori, we laid the plans for the first steps in an educational programme. It was to begin with the children in the Nursery from 1½ years upwards. To the opportunities already given these children for perfecting some of their grosser co-ordinations (in the gymnasium and in the infants' bath), it was proposed to add the brilliant apparatus devised by Montessori for the development of the finer co-ordinations, introducing into the Nursery an observer familiar with the use of this apparatus who would be willing to proceed along the general lines already laid down in the Centre. Those who are familiar with the Montessori apparatus, and particularly, for example, that devised for the appreciation of number, will know the illumination that can come even to the adult encountering it for the first time. Many people who by rule of thumb can manipulate cube root, have little sensuous appreciation of its significance. It was with the possibility of reawakening their own sense-perception that we planned to introduce young mothers one by one into the Nursery—so that in helping with the education of their own children not only might they come through practise to understand the essence of education, but that they might also simultaneously renew their own education long since forgotten and foregone.

Record of the spontaneous activities

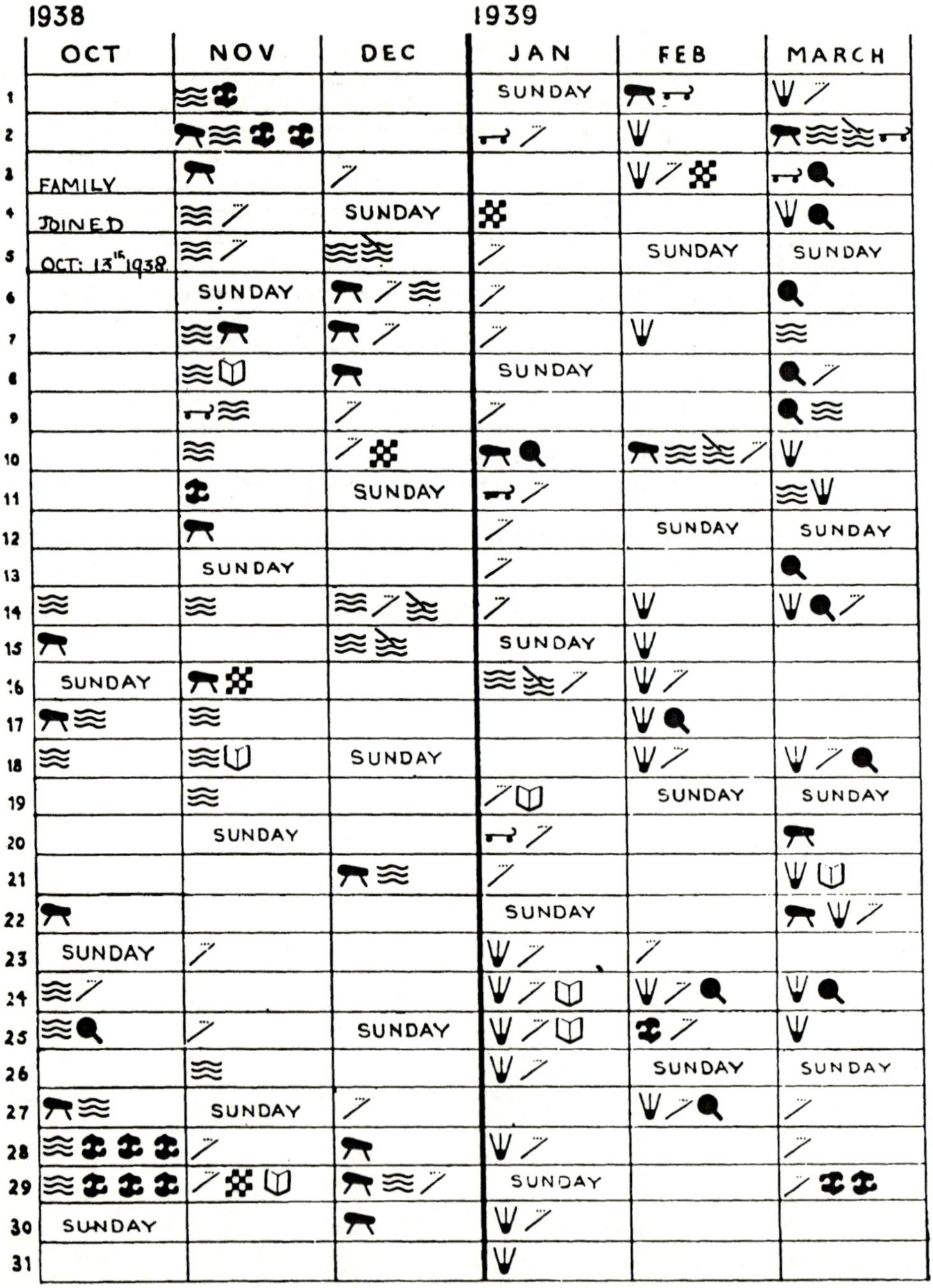

GYM: SWIMMING DIVING BADMINTON Table tennis BILLIARDS

N.B Two swims a day allowed in school holidays

in the Centre, of a boy (P M .) age 11

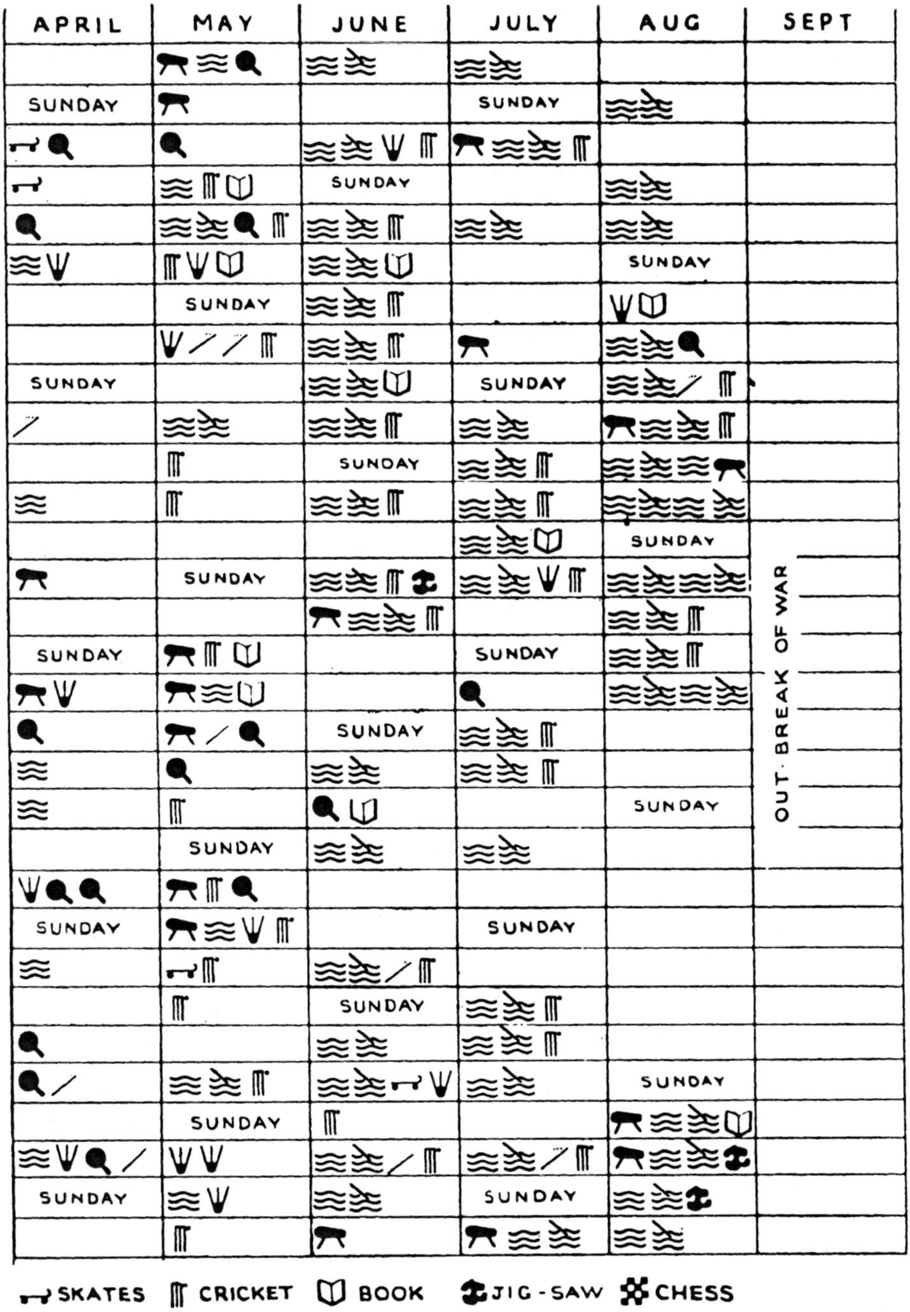

Financial and Administrative

This book would be incomplete without mention of the economics of a Health Centre. In this respect the Peckham Experiment, as its popular name implies, was no type-mechanism for all health centres. It was a frank experiment, and as such was built, equipped, and to a large degree maintained by the voluntary subscriptions of those anxious to see an attempt made to put into practice the principles it stood for. Moreover it was the first of its kind anywhere in the world—and as such had to weather the storm of strangeness amidst the populace for whom it was set up. How great is the effect of this factor of unfamiliarity in popularising any new invention can be judged from the history of the launching of many commercial concerns and products, the subsequent universal popularity of which cannot be questioned. Two examples will serve to illustrate this point—the failure of the first attempts to establish Woolworth stores, and the time—15 years—it took to popularise the use of anything so universally desirable as silk stockings.

This must lead us to suppose that acceptance of anything so new as a health centre must be a matter of time, however great its advantages may prove to those who understand its principles and to those who come within its doors and stay to experience its use. Already we have seen that in the case of family-membership in our chosen area there were many other factors—such for example as the absence of social aptitude, or the amount of moving that occurred in the district—that militated against stability of membership. As long as there are no Centres to integrate local society the first of these difficulties must persist, and until there are Centres in all districts so that a family which moves can be transferred complete with dossier to the Centre in the district to which it has migrated, the second must remain a serious source of loss. So even the collection and retention of members became for the first Centre a matter of experiment, putting off the date of the achievement of its full membership.

The Pioneer Health Centre was designed and built to cater for 2,000 member families (roughly 7,500 individuals). Experience has taught us that it was in every way well suited to fulfil this expectation. With the 1,200 people often within its doors on a Saturday night, there was still ample room for more, even before there had been time and opportunity to develop the extra-mural activities bound to emerge in any integrated society: e.g. camp, farm and special club activities.

Membership entailed the payment of a weekly subscription per family to include parents and all children of school age (1/- a week). A further sum was derivable from the adolescents and other adult individuals of a member-family, each of whom paid a further individual subscription (6d. a week). The proportion of these in the Centre was roughly one to every two member-families. Thus the income to accrue from full membership subscriptions forms a calculable one—roughly £6,000 a year.

Whereas all apparatus and equipment of the Centre was at the free disposal of children of school age and under, all adults paid a small sum for what they did in the Centre: e.g. billiards, the Keep Fit class, the Debating Society, use of the sewing machine, etc., as well as for all home remedies, substitutes for deficiencies (iron, vitamins, etc.), and emergency dressings supplied in the medical department and for all food bought in the cafeteria. This forms a further source of income, varying with the number of families and with the efficiency of the organisation, but calculated with a full membership to bring in a net sum of not less than the family membership subscriptions, namely, 1/- per week per family. These two sums together give an estimated income of over £10,000 per annum.

On the debit side, experience has shown us that running costs, excluding all research expenditure, amount to a sum approximating to £8,500 a year, and that it should be possible to keep them within £10,000 if payment of interest on capital[1] is included. It must be recalled that owing to the self-service basis of the organisation, the cost of running the Centre remains at much the same level whether it is catering for a membership of 500 or 2,000 families—the maximum charge, apart from the periodic overhaul, being that of lighting and heating the building, From our experience we must therefore infer that although the raising of the capital outlay for the establishment of a Centre may not come within the capacity of any local community, once built its *self-maintenance* is a matter of practical politics.

It is important to state that in our opinion self-maintenance of a Health Centre is not only a possibility, but is an *essential* for the maintenance of any institution *where health is the object*. The family's power to handle affairs, including the responsibility of adjusting a financial balance, is no less an expression of function—of the wholeness of apprehension of environmental circumstances—than any other capability. Moreover the balancing of a budget is one that the family with its limited means is very well competent to carry out, for it is no easy task to house, feed, clothe, educate, insure and maintain four or five people on the relatively slender weekly wage that is available for the purpose in the majority of working class homes.

Nevertheless the balancing of his budget is an experience from which, in the existing circumstances of life, the working man's family is cut out from all but his immediate domestic sphere. Even the slow steady increase in the level of wages does not alter this, for as wages rise, the tendency is for the State largely to determine for him on what that increase must be outlaid.

Management and knowledge of the family's social and local affairs in a place like the Centre is the next step in an education long overdue. The importance of some such expanding field of social activity coincident with a rising standard of wages is obvious for the development of health in the Nation.

In the case of *sickness* we are faced with a quite different proposition. The very essence of sickness is withdrawal from the environment—a

[1] say, of £50,000.

diminished power of the organism's spontaneous response to its circumstances. The burden of the doctor's work is, thus, to assume responsibility for the sufferer—to tell him what to do, how and when to do it—until he is again able to resume this responsibility for himself.

There is also yet another aspect of sickness. It is a liability and threat to society. It becomes therefore at once reasonable that the State itself should assume responsibility for the care of the invalid and for their speedy return to the ranks of the valid.

With the valid citizen it is different. The very definition of health implies an ability to work in *mutual* synthesis with the environment. The liberty to do so is the first right of the citizen. Without responsible excursion in its local society no Nation can hope to build up the capability for responsible and knowledgeable action required in a true Democracy.

Use of the Centre by its Members.

TABLE I shows the number of individuals entering the Centre day by day throughout one week (April 1938), and the times at which they arrived.

The approximate family membership at this date was 600 families.

Time of day	*Monday*	*Tuesday*	*Wednesday*	*Thursday*	*Friday*	*Saturday*
2— 3 p.m.	61	105	43	45	96	162
3— 4 p.m.	59	31	66	86	78	93
4— 5 p.m.	80	74	61	91	80	56
5— 6 p.m.	35	42	49	49	58	41
6— 7 p.m.	31	45	56	47	42	40
7— 8 p.m.	67	108	125	124	165	108
8— 9 p.m.	173	163	282	293	286	234
9—10 p.m.	76	81	59	70	83	227
Total	582	649	741	805	888	961

The above table has been compiled from a record of the numbers entering the building. As few individuals spent less than an hour at each visit to the Centre, and many spent three or four hours there at a stretch, the number of people in the building at any one time, after the first hour in the afternoon, naturally greatly exceeds those coming in during any one hour.

It should be noted that Thursday is early closing day for the local shops, Friday pay-day and that on Saturday the Centre is open till 11 p.m., instead of till 10.30 as on other evenings.

TABLE II shows an analysis of the individuals coming into the Centre during the course of one afternoon and evening.

Specimen day, Thursday, 24th November, 1938. Membership approximately 650 families.

Time of day	*Men*	*Women*	*School children*	*Infants*	*Total*
2— 3 p.m.	11	42	5	17	75
3— 4 p.m.	5	28	2	16	51
4— 5 p.m.	12	17	42	2	73
5— 6 p.m.	8	6	59	—	73
6— 7 p.m.	20	24	41	—	85
7— 8 p.m.	71	123	33	2	229
8— 9 p.m.	133	138	5	—	276
9—10 p.m.	67	81	1	—	149
Total	327	459	188	37	1011

V

Note :

1. The relatively large number of women coming in between 2 and 3 p.m., falling steadily to almost nil towards the time of the evening meal.
2. That the afternoon male population, small though it is, is somewhat above the average, since on a Thursday the local shops are shut and the men working in them at leisure.
3. The high proportion of school children between 4 and 8 p.m.
4. The predominately adult population from 8 p.m. onwards.
5. The late arrival of men, indicative of their restricted leisure available for family life.

INDEX

All sub-sections dealing with a particular subject are indicated by numerals in heavy type, definitions and references of outstanding importance in italics

accidents, 181, 185.
action, 43, 47, 79, 113, 114, 120, 130, 139, 196, *245*, 257, 272, *274*, 279, 280.
——— appetite for, 182.
——— chances for, 44, 79, 138, 229, 268, 289.
——— incentive to, **126-27**, 181, 185, 196, 270.
——— limitation of, 105, 138, 258, 266, 267, *see also* 121, 165.
——— spontaneous, 69, 125, 128, 201, *218*. *See also* spontaneity.
action-pattern,' 27, *131*, 200, 209, 211, 224, 227, 238, 240, 247, 273, 298.
activities, 68, 74, 126-30, 196, 201, 209-211, 219, 267, 269, *278*, 288, 296, 301, **303-306**.
——— chart of child's, **318-319**.
——— initiation of, *129-130*, 199.
activity, zones of, 204, 214.
administration, 245, 277, 281, **320-322**.
adolescence, 45, 144, 206, **207-223**, 229-232, 244, *268-71*, 280, 289, 294-295, 297.
after-care, 278, 284-285. *See also* 94, 106.
agriculture, 24, 147-148, 271.
allergy, 87, 115, 188, 234, 241.
almoners, 251, 278.
altruism, 'physical', *181*, 193, 213, 234, 249.
amoeba, 21-22, 163.
ante-natal care, 110, **139-57**, 278, 280-281, 303.
appetitive phases, *171*, 177-8, *182-183*, 196, 198, 201, 2 4, 217, 229, 232, 251, 256, 272.
appointments for overhaul, 80-83, 139, 277.
authority, 78, 270.
autonomic (control), *28*, 32, 38-40, 45, 225-6.
average, *see* 'normal'.
axylotl, 20, 45.

Balfour, Lady Eve, 24.
'bias' (of sex), 208, 225.
bee, 17, 32, 131, 240.
— queen, 17, 32.
'biological junctions', **243-245**, *see also* 144.
biologist, the, 15, 21-2, 26, 40, 46, 77, 84, 90, 92, 107, **111-113**, 123, 136, 143, 147, 153, 159, 162, 171, 183, 212-3, 245, 276, 280.
bi-polarity (sex), *18-19*, 163, 208, 224.
birth, **162-174**, 281, 282.
—— control, *see* conception, control of.
—— of a family, 228, 236, **237-246**.
birth rate, 138, **258-259**.
breast feeding, *see* feeding.
breast milk, *see* milk, breast.
Bridges, Robert, 17.
building, the, **67-70**, 71, 126, 127, 130, **301-302**.
——— plans of, 300.
——— heating of, 301.
budget, balancing of family, 321.

camp, the, 71, 133, 219, 296, **305-306**.
cancer, 25, 97, 101, 102, 111, 112, 114.
'centripetal phase', 34, *166-167*, 172, 245.
chastity, 233-234.
child, difficult, 284.
—— motherbound (or skirt-bound), 183, 186, *261-263*, 290.
—— only, 181, 185, 255.
—— spoilt, 173, 199, 260.
Child Guidance, 251, 278, *283-284*.
Child Welfare, 11, 97, 162, 244, 257, 278, 280, *281-283*, 303. *See also* Infant Consultations.
chronicity, *100-101*, 105, 107, 279 282.
clinical material, 277.

clinical medicine, *see* medicine, clinical.
clinician, the, 95, 105-107, 108, 111, 112, 143, 154, 244, 281, 284.
clinics, 111, 281, *292.*
clubs, intramural, 125, **127-130.**
coleoptile, *see* 'growing point'.
committee(s), 127-130, 295.
community, *291-292*, 297, 298.
compensation, *103-107*, 115, 138, 145, 178, 183, 273.
compensative existence, *103-105*, 244, 252-253, 258, 262-264. *See also* well-being.
competition, 195.
conception, 11, 88, **135-139**, 145, 153, 155, 162, 168, 260.
———— control of, *89*, 258, 260, 278, 303.
conditioned reflexes, 193.
confinement', 155, 157, 165-166, 283.
consultations, family, 81, **84-92**, 93, 109, 112, 137, 186, 212, 237, 247, 251, 253, 254, 258, 265, 303.
———— infant, *see* infant consultations.
———— 'parental', 94, 104, *148-151*, 168-169, 175, 303.
———— 'premarital', *237-238*, *241-243*, 303.
continuity of association, *44*, 229.
convalescence, 278, *284-285*. *See also* 106.
cookery classes, 256.
co-ordination, 22, 24, 114, *119-121*, 139, 153, 159-160, *170-173*, 178, 182-184, 201, 206, 267.
cost of building, 301.
— of maintenance, 74, 321.
— of overhaul, **116-117**, **313-314.**
courtship, 33, 86, 144, 153, 221, 223, **224-236**, 244, *272*, 280.
cultivator, 23, *123*, 141, 236, 244.
Curator, the, *78*, 80, 124, 127, 129, 197-199, 212, 216, 276.

'dancing class', 216-218.
Darwin, Charles, 40.
deficiencies, 107, 108, 138, 140-148, 153, 155, 159, 193, 228, 247, 254, 271.
delivery, 89, 150, **153-159**, 165, 166.
development, 14, 25, 28, 30, 34, 39, 44, 45, 88, 92, 105, 131, 136, 138, *183*, 184, 196, 202, 208, 209, 212, 215, 220, 229, 236, 249, 267, 273, 279, 283, 286, 289, 292.
———— arrested, 245, 273.
devitalisation, 12, 107, 138, 182, 231, 258-261, 271, 290.
diagnosis, early, 97, 101, 106, 107, **108-110**, 111, 117, 275, 292.
———— of health, 111, 112.
diathesis, 115, 146, 157, 178, 244, 252.
diet, 109, 112, 122, 145-150, 176-179.
dietery substitution, 142, *144-147*, 159, 258.
differentiation, 22, 28, 139, 211, 222, 227, 229, *243-245*, 281.
digestion, 87, 170, 173, 176-178.
———— of experience, 90, 122, 159, 188, 239, 241, 246, 264.
discipline, 41, *200*, 270.
———— in the observer, 46.
disease, 13, 21, **96-101**, 111, 251, 254, 262, 273.
discrimination, 163, 200, 202, 204, 223, 225, 226, 229-231, 280.
disorder(s), 10, 13, 21, *95-97*, 101, 105, 111, 138, 140, 146, 155, 159, 178, 228, 247, 251, 254, 262, 272, 273, 275, 279, 284.
disorder, effect of removal of, **113-116.**
district, the, **70-72**, 73.
diversity, *24*, *41*, *43-46*, 71, 131 145, 192, 207, 219, 222, 227, 232, 233.
———— cultural, 44, 46, 71.
diversification, *24*, 25, 27, 227, 229, 241.
doves, 37, 88.
'dowry', 30, 32.

dowry of skill, 255.
'drugs', 147, 271, 290.
Durden, Field and Percy Smith, 21.

Ecology, *see* Oecology.
education, 11, 121, 122, 171, 182, 188, 192, 193, 194, 199, 200, 202, 209, 220, 221, 223, 245, 255, 256, 267, 277, 278, 280 286, 321.
——— experiment in, **317.**
——— of married woman, 283, 286.
——— science of, *183.*
——— topical, *179*, 217.
educative principle, 195.
egotism, 181, 192, 193, 207, 235, 249.
endocrine (secretions), 20, 29, 115, 153, 170, 225, 232, 233, 256, 260.
endocrinology, 136, 156, 164, 258.
embryo, the, 31, 34 et seq., 75, 141.
embryology, 25, 183.
employment, nature of members, **307-310.**
enrolment, **80-82**, 124.
environment, 12, 16, *21-26*, 30, 35, 39, 41, 43-47, 103-105, 107, 113, 116, 120-123, 138-139, 145, 162-164, 174, 182, 183, 185, 187, 202, 218, 227, 229, 232, 239-242, 243, 250, 258, 261, 265, 273, 279, 289, *290-291.*
——— 'hostility' of, 22, 24.
——— social, 120, 270. :
equipment, 68, 72, 75, 79 283
——— children's access to, 197-199.
'establishment', *171*, 172, 173, 183, 221.
evolution, *25*, 35, 40, 226.
eugenics, 277.

facts, *see* knowledge
facultization, *163*, 182, 193-195, 196, 202, 206, *208*, 226, 232, 272.
——— of sex, *221-223.*

familiar nurture, *see* nurture, familiar.
familiarisation, *36-39*, 70, 72, 143, 155, 169, 175, 233, 241, 290, 291.
family, the, 9, *20*, 27, 28, 39, 41, 42-48, 70, 72, 84, *92*, 108, 120-123, 140, 144, 162-168, *172*, 175, 185, 189, 225, 228, 236, 237, 239, 241, 278, 281, 286-287, 289-292, *297-298.*
——— as a biological mechanism, 188, 189, 213, 281. *See also* 204, 205.
——— club, 11, **72**, 279.
——— consultation, *see* consultation, family.
——— membership, 71-73, 230.
——— practitioner, 108, 110.
farm, the, 147, *148*, 150, 179, 219, 295, **305.**
fear, 83, 85, 152, *155*, 158, 177, 184, 186, 191, 212, 227, 232, 248, 268, 285.
feeding, artificial, 169.
——— breast, 36, 38, 87, 157, 158, 160, 165, 167, *169-174*, 282.
finance, 73, 74, 280, 295, **320-322.** *See also* costs.
financial stability of members, 310.
Flack, Martin, 119.
food (nutriment), *22-24*, 30, 34-39, 45, 87, 147, 151, 172, 175-178, 189, 207, 212, 239-241, 246, 250, 263, 271, 273, 289.
——— quality of, *31*, 146, 148, 149-150, 179.
friends, 87, 135, 174, 175, 184, 205, 207, 219, 251, 266, 285, 286.
friendlessness, 131, 248, 249, 257, 261.
function, *15-26*, 35, 92, 115, 116, 117-123, 153, 159, 160, 162, 171, 178, 185, 193, 247, 259, 273, 289, *298.*
functional organisation, 18, 34, 35, *38*, 41, 120, 292, 297.

genetics, 120, 144, 226, 227, 272.
goodwill, 274.
——— of parents, *149*, 213.

growing point (coleoptile), 82, *163*, 164, 179, 188, 239, 240, 250. *See also* 203.
growth, 243. *See also* development.
gymnasium, 67, 69, 132, 180, *181-185*, *191-193*, 200, 293.
——— floor, 302.

Haldane, J. S., 28.
health, 9-11, 21, *24*, 26, 42, 86, 101, 105-107, 113, **117-123**, 137, 147, 157, 158, 177, 178, 221, 225, 226, 232, 235, 237, 240, 241, 243, 244, 247, 267, 271, 321-322.
——— cultivation of, 12, 69, 101, 05, 120, 144, 153, *229*, *236*, 245, 275, 276, 277, 281, 283.
——— infectious, 274.
——— measurement of, *see* standards.
——— practice of, 14, 106, 113, 144, *246*, 277.
Health Centre, cultural, 12, 13, 48, 110, 116, 276, *292*.
'health centres', 49, 292.
health overhaul, *see* periodic health overhaul.
——— after sickness, 94, 106-107, 284.
hearth, *see* 'nest'.
home, 87, 88, 164, 179, 204, 207, 208, 232, 236, **238-243**, 246, 247-248, 250, 251, 252, 263, 265, 268, 287, 289, 291, 292, *297-298*.
homoculture, 91.
husband, extrusion of from family, 263.
hospital, 10, 107, 111, 113, 156, 157, 278, 285, 296.
house, the, 43, 165, 239, 242, 248, 249, 250, 255, 257, 259.
housing, 41, 70, 156, 158.
Howard, Sir Albert, 148.
Huxley, Julian, 228.
Hygiene, 10, 11, 296.
hypotonia, 258.
hysteria of repression, 199-201.

immaturity, 20, 215, 220, 222, 223, 295.
——— of parents, 270.
immunisation, 173, 278, *285-286*, 303.
immunology, 233, 237. *See also 10*.
inbreeding, 227, 228, 272.
individuality, 25, *27*, 28, 31, 163, 167, *208*, 210-212, 234.
Indore (compost) method, 148.
Industry, 41, 43, 101, *213*, 220, 221.
infant, *11*, *162-179*, 281-283, 289.
——— consultations, 93, 168, 172, 176, 178, 303.
——— welfare, *see* Child Welfare.
infertility, 137, *259*.
information, the giving of, *see* knowledge.
instinct, 177, 180, 225-227, 257, 272.
instinctive behaviour, *see* chap. II.
instructors, *see* teachers.
instruments, educational, 124, *194-199*, 202, 203, 290.
——— of health, *12*, 77, 78, 92, 280, 281, 283-285.
——— self-evident, 195, 196.
——— of therapy, *276*.
intrepidity, *119-121*, 206.
iron deficiency, 107, 114, 141-145, 149, 200, **315**.

Jeans, Sir James, 15.

'knapsack', 30, 31, 206.
knowledge, 43, 47, 79, 90, 92, 108, 109, 136, 139, 149, 168, 175, 176, 179, 180, 184, 192, 195, 212, 213, 217, 221, 229, 242, 245, 246, 253, 256, *274*, *279*, 280, 284, 285, 286, 287, 289, 290. *See also* 152.
key, the, **76-77**, 125.
Kirkman, F. B., 164.

laboratory, the biologist's, 43, *48*, 68.
——— examination, 81-83, 109, 112, 117, 139, 141, 145, 151, 158, 168, 212, 283, **311**.
labour, *see* 'delivery'.
lactation, *see* 'feeding'.
lactagogue, natural, 174.

language, common basis of, 85, 91.
leadership, 69, 120, 128, 129, 214, **270-271**, 274.
legal advice, 278, 303.
leisure, *43*, 46, 48, 71, 187, 189, 219, 221, 222, 241, 255, 256, 279, 280, 284, 287, 288.
——— of father, 203, 205, 254.
love, falling in, 144, *225-235*, 271.

maintenance (of Centre), 73-76, 280, 321.
Marais, E., 17.
marriage, 18, 144, 234, **237-239**, 244, 263.
——— early days of, 87, 239, 241, 242, **245-246**, 251, 252, 255-257.
——— unconsummated, 137, 257-259.
Maternity practice of biologist, 153-157. *See also* ante-natal care.
——— service, 97, 110, 156, *161*, 244, 281.
mating, 18, 36, 41, 163, 223, 225-229, 231-236, 237-238, 241-244, 251, 271, 272, 280.
maturation, 29, 38, 223, 229, 230, 233, 242.
maturity, 44, 46, 88, 139, 170, 172, 185, 188, 210, 215, 220,.222, 223, 237, 246, 270.
Medicine, 14, 77, 93, 100, 113, 277.
——— clinical, 101, 104, 107, 115, 117.
——— science of, 94, 97, 101, 103, 106, 107, 110, 117.
——— student of, 26, 296.
Medical Planning Commission Interim Report, 49.
medical profession, 244, 259.
Medical Services, 97, 111, 117.
membership, attendance, **323**.
members, temporary, 230, 303.
metamorphosis, 20, 45, 232, 234, 237, 244.
midwife, 139, 151, 157, 158, 276.
milk, breast, 38, 39, 143, 169, 171, 175, 188.
—— cows', 146, 147, 179, 295, 305.
—— crop, 37.
Montessori, Maria, 182, 317.
Morris, J. N., 95.
mutual (action), 18, 25, 92, 125, 155, 193, *203*, 224, 264.
mutual synthesis, *24*, 26, 35-39, 42, 44-45, 103, 105, 107, 122, 142, 154-155, 163, 169, 195, 202, 213, 232, 234-235, 239, 241, *292*, *298*, 322.
mutuality, *23-26*, 31, 33-35, 136, 172, 185, 208, 219, 233, 238, 263, 297.
——— zone of, *35-36*, 39, *164*, 204-205, *291-292*.

nest (or hearth), 29, 30, 31, 35, 37, 38, 122, *155-157*, 164-167, 172, 179, 181, 204, 239, 247, 248, 282, 297.
nesting, 35, 140, 153, 162, 164, *165*, 166, *235*, *236*, 257, 282.
nidation, *see* nesting.
nidus, *see* nest.
noise, 301.
'normal', 10, 120, 143, 161, 162.
novelty, 19, *27*, 131, 163, 235, 238.
nursery, the Centre, 67, 167, 172, 261, 285, 300, 317.
——— afternoon, 152, *180-185*, 186, 252, 261, 262, 292, 304.
——— night, 174, 221, 304.
——— staff of, 152, 276.
nursery teas, 178, 179, *256-257*, 293, 304.
Nurseries, communal, 292.
——— therapeutic and preventive, 283, 278.
nurture, 120, 186, *188*, 202, 203, *227*, *272*, 286.
——— ebb of instinct for, **254-258**.
——— 'familiar', *31-39*, *45*, 48, 121, 122, 144, 145, 163, 177, 208, 232, 264.
nutriment, *see* 'food'.
nutrition, 108, 122, *145-148*.
nutritional surveys, 118.

obstetrician, the, 154, 156, 157, 259, 283.
Oecology, 24, 131.

opportunity, *45*, 168, 173, *186*, 205, 223, 228, 232, 236, 242, 254, 262, 274.
order, *25*, *26*, 128, 130, 211.
—— 'in anarchy', 130.
—— social, 181, *199*, 284. *See also* 298.
'organ-ation' of the environment, *25*.
organism, *16*, *17*, 24, 26, 37, 41, 103, 131, 144, 162-163, *228*, *237*, 239, 240, 289.
overclothing, 172.
ovum (egg), *28-35*, 37, 172, 233, 243, 290.

packs, adolescent, 210-215.
parental immaturity, 270.
parenthood, *19*, 36-39, 45, *86*, 145, 162, 166, 208, 235-236, 238-244, 248, 258, 289, 297-298.
—— atrophic, 268.
—— inco-ordination of, 72.
pathologist, the, 10, 25, *245*.
pathology, *9-11*, 21, 26, 42, 46, *93*, 160, 225, 228, 247, 251, 270, 272, 273, 274, 298.
—— social, *273*.
patient, the, 10, *95*, 97, 110, 155-156, 159-160.
periodic health overhaul, 11, *12*, 72, 77, **79-92**, 93, 95, 100, 102, *107*, 110, 112, *116*, 117, 137, 140, 157, 168, 230, 253, 254, 264, 265, 277, 280, 284, 303.
—— cost of, 116, 117, **313-314.**
periodic medical overhaul, 72, 81, 100, *116*.
phase, centripetal, *see* 'centripetal phase'.
phase, plastic (or formative), *33*, 136, 144, 149, 168, *243*.
—— refractory, *178*, 251.
physiological v. functional, *15-16*, 26, 120, 121, 193.
placenta, *35-36*, 38, 140, 154, 162, 172, 236.
—— social, 204-205, 291.
Plesch 'tonosillogram', 142, 312.
Polyclinics, 49, 292.
'pool of information', *79*, 90-92, 242.
pregnancy, 28, *32*, *35*, 38, 87-89, 94, 103, 137, **139-159**, 162, 168, 176, *237*, 242, *243-244*, 250, 253, 258, 259.
—— husband's attitude to, *150*.
—— unwanted, 137, 245, 260.
—— wanted, *136*, *137*.
preventive measures, 109, 283, 285.
prism, 13, 117.
professionalism, 86, 127, **266-267.**
promiscuity, 230, 234.
psyche, birth of the, 208, 211.
psychology, 46, 144, 213, 232, 238, 269, **272-273**, 277.
psycho-pathology, 104, 168, 231, 251, *272-273*, 274, 290.
puberty, 207-210, 212, 231, 243. *See also* adolescence.
puerperium, **159-161**, 170, 282.

quality, *19*, 27, *31*, 37, 131, 154, 167, 169, 234.

'raw material', 86, 246, 279, 290. *See also* 131.
reaction (pathological), 207, 235, 247, 259, 270, 272, 286.
records (dossier), 76, 110, 198, 283-284. *See also* 141, 157.
recoverability, 249.
'refractory phase', *178*, 251.
rehabilitation, 145, 278, 284-285.
reproduction, 19, *20*, 27.
research (study or experiment), 10, 12, 26, 36, 40-48, 76, 111, 114, 115, 117, 144, 170, 171, 178, 195, 203, 212, 213, 220, 228, 272-273, 277, 283, 313, 314, 321.

reserves, physiological, 21, *103-107*, 113, 141, 145-146, 150, *153-155*, 159, 253, 259, 284.
——— 'floating', 142-144.
responsibility (responsible action), 75, 177, 220, 279, 280, 286, *321-322*.
revenue (income), 73, 74, 125, 320-321.
Riddle, O., 37.
rules and regulations, absence of, 86, 128, 196.

salaries, staff, 76, 313, 314.
scholarship boy, 289.
school, 188, 189, 200, 202, 221, 241, 289, **317**.
schoolchild, **188-206**, 220, 264, 294.
——— medical inspection of, 94, 97, 117, 118, 278, 280.
segregation, adolescent, *223*, 270.
——— sex, 215, 270, 272.
——— in society, *28*, 44, 215, 272.
——— wage-levels, 47, 272.
self-maintenance (of Centre), 73, 74, 280, *321*.
self-service, **74-76**, 78, 130, 295.
sensitization, 23, 31, 34, 154, 167, 233. *See also* 237, 286-287.
sex, *18-20*, 208, 211, 214, 215, 220-223, *224-236*, 237, 238, 258-259.
— education, 224, 303.
sexes, equality of the, 225.
—— segregation of the, 215, 270, 272.
Sherrington, Sir Charles, 24.
sick, the, 117, 157, 281, 282.
—— Maternity Service for, *161*.
sickness, *9*, *10*, 12, 21, 93, *101*, *107*, 108, 111, *113*, 154, 273, *274*, 284, *321-322*.
——— chronicity of, *100*, 105.
——— prevention of, 109.
——— 'morning', 140.
skill, *126*, 128-130, 194, *195*, 197, 199, 200, 204, 214, 215, 217, 219, 221, *255*, 256, 264-265, *266-267* 270, 293, 296.
skill in diagnosis and treatment, 110, 112, 117, 285.
skirt weaning, *see* weaning.
social, 284, *290-291*.
—— action, 172, 185, 196, 248, 258, 261.
—— —— centripetal phase of, 166-167, 245.
—— competance (co-ordination), 206, 215, 221.
—— contacts, *41*, *43*, *69*, 134, 151, 174, 238, 249, 272.
—— education, 221, 286.
—— faculties, 136, 139.
—— integration, *121-122*, *125*, 206, 232, 272, 286, 288, *290*, *291*, *296*.
—— isolation, 148, 260, 261, 264, 289.
—— life, 185, 269, 290.
—— pathology, *273*.
—— problem class, 42, 94, 159, 261. *See also* 310.
—— remedy, **265-266**, 271
—— services, 271.
—— stagnation, 254, 271.
—— starvation, 207, 248, 254.
—— worker, 46, 244, 251, 271.
society, 131, 232, 239-240, 264, 271, *291-292*, 296, *297*.
——— adult (children in), 197.
——— of the Centre, 197, 228, 246, 264, 267, 272, 286.
——— conditions in, 166, 227, 228, 235, 244, 251, 254.
——— disintegration of, 207, 211, 212, 247, *271-273*, 298, *see also* 223.
——— drug sustained, 271. *See also* 290.
——— idiom of a, 222.
——— inco-ordination in, 192-193.
——— integration of, *see* social integration.
——— living, *9*, 291, *298*. *See also* 291.
——— mixed, 270, 284.
——— sample of, 42, 43.
——— unfacultised, 267.
——— vertical grouping in, 44, 228.
soil, 23, 131, *144*, 147, **163**, 239, *244*, *246*, *249*.

soil, social (milieu), 119-120, *123*, 221, 229, 236, 243, 254, *271*.
specialist, the, *39*, 126, 130, 267, 272.
——— educational, 188.
——— medical, 283.
specificity, 18, *23*, *24*, 31-33. 43, 87, 145, 169, 177-178, 229, 233, 234, 236, 239-240, 298.
sperm, 29, 32, 233.
spontaneity, 69, 105, 119, *128*, 130-131, *193*, 214, 230, 290.
staff, the, **77-78**, 79, 83, 90, 198, 199, 213, 268, 275, *276-277*, 279, 292, 294, 296, *313*, *314*.
—— training of, 46, 78, 110.
standard of living, *138*, *see also* 254-255.
standards (measurements), *117-118*.
——— clinical, 118. *See also* 100.
——— functional (biological), 117, *118-123*, 141-147, 183.
——— health, *115-116*, *117*, *118*. *See also* 112-113.
——— physiological, 118. *See also* 159.
statistical data, 41, 76, 114, 246.
stimuli, summation of, 160, 223.
stimulus, 126, 189, 222-223, 270-271, 290. *See also* 255.
stools (infants'), 158, 165, 176.
student, the, 198, 199, 276, 292, 293.
——— of health, *17*, *18*, 20, *26*, 113.
——— medical, *see* Medicine.
subscription (membership), *72*, 76, 82, 86, *230*, 286, *303*, 320.
swimming bath, 67, 69-70, 126-127, 132-133, 189, 205, 221, 222, 293, *300*, *301*, 304.
——— learners', 180, 184, 294.
synthesis, *23-24*, *27*, 28, 236. *See also* mutual synthesis.

taboo, 227.
tadpole, 20, 237.
teacher (instructor), 86, 194, 196, 199, 202, 204, 216-217.
teaching, outlook on, 257, **266-267**.
——— talent for, 217.
technique, new, *48*, 74, *91-92*, 245, 276.
territory, inner of birds, 164, 235. *See also* intimate environment, 46.
therapist, the, 276, 281, 290.
therapeutic agencies, 275, 276.
therapy, 106, 147, 177, 193-194.
tickets, *197-198*, 205, 294.
topicality, *45*, 92, 179, 212, 217, 286, 287.
training, 120, 122, 128, *193-194*.
trampolin, 62, 195.
treatment, 97, 100-101, 106, 109, 275, 283, 284.
——— of the early case, **110-111**, 276.
'two-eyed vision', 19, 235.

unemployment, 273, 292, 310.
unit (of function), *9*, *16*, *20*, 40, 292, *298*.
unity, *18-19*, 29, *32-33*, 41, 86, 163, *234-236*, 237, 240, 265, 291.
U.S.A. Senate Document No. 3489, 95.
utilization, *45*, 122, 139, 143, *146-147*, 150, 159, 188, *244*, 291.

vaccination, 173, 286.
visibility (in building), *68*, 70.
vitamin deficiency, 102, 107, 114, 200.
——— substitution, 115, 147.
vocational guidance, *213*, 278, 280, 303. *See also* 90.
voluntary (gnomic) wisdom, 28, 226.
wage-earner, 82, 231.
wages, wage level, *70*, 252, 254, 255-257, 273, *274*, 280, *310*, *321*.
weaning, *90*, 188, 251, 256.
——— birth, **162-174**, 208.
——— breast, 90, 143, **174-179**.
——— skirt, *90*, 168, **180-187**, 261-262, 290.
——— from nest, *see* 207, 208, 268.
Weiss, Paul, 30, 33.
well-being, *13*, 21, *96*, *98*, *99*, **101-117**. *See also* compensative existence.

Williams, Sir E. Owen, 301.
Williamson, G. Scott, 5, 77, 301, 302.
wooing, 33, *233-234*.
worms, 102, 114, 146, 200, 231, 249, 252, **315-316**.
Wrench, G. T. 175.

Youth, **268-271**.
Youth Organisations, 215, 270.

zone of mutuality, *see* mutuality.